Lecture Notes in Physics

New Series m: Monographs

The Editorial Policy for Monographs

The series Lecture Notes in Physics reports new developments in physical research and teaching - quickly, informally, and at a high level. The type of material considered for publication in the New Series m includes monographs presenting original research or new angles in a classical field. The timeliness of a manuscript is more important than its form, which may be preliminary or tentative. Manuscripts should be reasonably self-contained. They will often present not only results of the author(s) but also related work by other people and will provide sufficient motivation, examples, and applications.
The manuscripts or a detailed description thereof should be submitted either to one of the series editors or to the managing editor. The proposal is then carefully refereed. A final decision concerning publication can often only be made on the basis of the complete manuscript, but otherwise the editors will try to make a preliminary decision as definite as they can on the basis of the available information.
Manuscripts should be no less than 100 and preferably no more than 400 pages in length. Final manuscripts should preferably be in English, or possibly in French or German. They should include a table of contents and an informative introduction accessible also to readers not particularly familiar with the topic treated. Authors are free to use the material in other publications. However, if extensive use is made elsewhere, the publisher should be informed. Authors receive jointly 50 complimentary copies of their book. They are entitled to purchase further copies of their book at a reduced rate. As a rule no reprints of individual contributions can be supplied. No royalty is paid on Lecture Notes in Physics volumes. Commitment to publish is made by letter of interest rather than by signing a formal contract. Springer-Verlag secures the copyright for each volume.

The Production Process

The books are hardbound, and quality paper appropriate to the needs of the author(s) is used. Publication time is about ten weeks. More than twenty years of experience guarantee authors the best possible service. To reach the goal of rapid publication at a low price the technique of photographic reproduction from a camera-ready manuscript was chosen. This process shifts the main responsibility for the technical quality considerably from the publisher to the author. We therefore urge all authors to observe very carefully our guidelines for the preparation of camera-ready manuscripts, which we will supply on request. This applies especially to the quality of figures and halftones submitted for publication. Figures should be submitted as originals or glossy prints, as very often Xerox copies are not suitable for reproduction. In addition, it might be useful to look at some of the volumes already published or, especially if some atypical text is planned, to write to the Physics Editorial Department of Springer-Verlag direct. This avoids mistakes and time-consuming correspondence during the production period.
As a special service, we offer free of charge LaTeX and TeX macro packages to format the text according to Springer-Verlag's quality requirements. We strongly recommend authors to make use of this offer, as the result will be a book of considerably improved technical quality. The typescript will be reduced in size (75% of the original). Therefore, for example, any writing within figures should not be smaller than 2.5 mm.
Manuscripts not meeting the technical standard of the series will have to be returned for improvement.
For further information please contact Springer-Verlag, Physics Editorial Department II, Tiergartenstrasse 17, W-6900 Heidelberg, FRG.

Howard Carmichael

An Open Systems Approach to Quantum Optics

Lectures Presented at
the Université Libre de Bruxelles
October 28 to November 4, 1991

Springer-Verlag Berlin Heidelberg GmbH

Author

Howard Carmichael
University of Oregon, Department of Physics
College of Arts and Sciences, 120 Willamette Hall
Eugene, OR 97403-1274, USA

ISBN 978-3-662-13926-4 ISBN 978-3-540-47620-7 (eBook)
DOI 10.1007/978-3-540-47620-7

Originally published by Springer-Verlag Berlin Heidelberg New York in 1993

2158/3140-543210

Dedication

To Marybeth
and the little Welshmen

Acknowledgements

I would like to thank Professor Paul Mandel, Optique Nonlinéaire Théorique, Université Libre de Bruxelles, for inviting me to give the *ULB Lectures in Nonlinear Optics* and for his hospitality to me and my wife during our time in Brussels. I thank the Université Libre de Bruxelles for an appointment as Visiting Professor during my stay, with financial support from the Interuniversity Attraction Pole program of the Belgian government. The early part of this volume relies heavily on material developed from lectures I gave while a Visiting Lecturer at the University of Texas at Austin during the fall semester of 1984. I thank Professor Jeff Kimble, now at the California Institute of Technology, for hosting me on that occasion, and for the benefits of subsequent scientific interactions with himself and his group. Liguang Tian has been an invaluable help with the work on quantum trajectories discussed in the latter part of the volume. She is responsible for all of the numerical simulations and for preparing many of the figures. We have learned a great deal about quantum trajectories since she began working with the idea two and a half years ago. I would also like to acknowledge Murray Wolinsky and Phil Kochan who have worked as students on aspects of the theory. Murray was able to produce our first quantum trajectory simulations using a density operator formulation of the theory after a brief "napkin discussion" in a Chinese restaurant in 1989. Regrettably, the work of all of these students has received less exposure than ideally I – they, I am sure – would like.

Preface

This volume contains ten lectures presented in the series *ULB Lectures in Nonlinear Optics* at the Université Libre de Bruxelles during the period October 28 to November 4, 1991. A large part of the first six lectures is taken from material prepared for a book of somewhat larger scope which will be published by Springer under the title *Quantum Statistical Methods in Quantum Optics*. The principal reason for the early publication of the present volume concerns the material contained in the last four lectures. Here I have put together, in a more or less systematic way, some ideas about the use of stochastic wavefunctions in the theory of open quantum optical systems. These ideas were developed with the help of two of my students, Murray Wolinsky and Liguang Tian, over a period of approximately two years. They are built on a foundation laid down in a paper written with Surendra Singh, Reeta Vyas, and Perry Rice on waiting-time distributions and wavefunction collapse in resonance fluorescence [*Phys. Rev.* A, **39**, 1200 (1989)]. The ULB lecture notes contain my first serious attempt to give a complete account of the ideas and their potential applications. I am grateful to Professor Paul Mandel who, through his invitation to give the lectures, stimulated me to organize something useful out of work that may, otherwise, have waited considerably longer to be brought together.

At this time, more than a year after I presented the ULB Lectures, the account in this volume is far from complete. I have continued my work with Liguang Tian and a new student, Phil Kochan, and now there is quite a lot more we could say. More important than this, there is related work by other people, none of which is referenced in the lectures. The related work falls into two categories: work in quantum optics, some of it published nearly simultaneously with my lectures and some published during the last year, and work from outside quantum optics coming, in the main, from measurement theory circles. The second category includes work that predates my lectures by a number of years. I would like to give a full account of everything and, in particular, comment on the relationships between the ideas in this volume and those coming from measurement theory and elsewhere. It is not practical, however, to attempt this and still see the ULB Lectures published in a reasonable time; there has already been a long delay due to my late realization that I should publish the lectures ahead of the larger book I am writing. As a partial solution I have added a postscript that lists the most relevant references I am aware of; in the postscript I make some brief comments that are intended only to classify the references in a very general way.

Eugene, Oregon
January 1993

H. J. Carmichael

References

[1] I. R. Senitzky, *Phys. Rev.*, **119**, 670 (1960); **124**, 642 (1961).
[2] R. J. Glauber, *Phys. Rev. Lett.* **10**, 84 (1963).
[3] R. J. Glauber, *Phys. Rev.* **130**, 2529 (1963).
[4] R. J. Glauber, *Phys. Rev.* **131**, 2766 (1963).
[5] E. C. G. Sudarshan, *Phys. Rev. Lett.* **10**, 277 (1963).
[6] B. R. Mollow and R. J. Glauber: Phys. Rev. **160**, 1076 (1967); **160**, 1097 (1967).
[7] J. R. Klauder, J. McKenna, and D. G. Currie, *J. Math. Phys.* **6**, 734 (1965).
[8] C. L. Mehta and E. C. G. Sudarshan, *Phys. Rev.* **138**, B274 (1965).
[9] J. R. Klauder, *Phys. Rev. Lett.* **16**, 534 (1966).

like this the methods for turning master equations into Fokker-Planck equations are generally not useful.

The states of the optical field accessed through nonlinear interactions have been brought to our attention through the study of antibunched, sub-Poissonian, and most recently, quadrature squeezed light. Squeezing, certainly, is the principal popularizer. The theoretical challenges posed by squeezing did not, however, strain standard methodologies too far, because for practical systems the nonlinear interactions involved in squeezing generate transformations that combine field modes in a linear way. As a consequence, the most transparent calculations are operator based, with the open systems character removed by transforming the modes of a closed system, two at a time, in frequency space.

The more telling strain on standard methods for dealing with master equations has come from another direction – from the field of cavity quantum electrodynamics (cavity Q.E.D.). In cavity Q.E.D. the many-photon difficulty of optical systems that Glauber dealt with is, in one sense, removed. The emphasis moves closer to standard Q.E.D.. The whole objective in cavity Q.E.D. is to achieve experimental conditions that make the field of just one photon large, where "large" is measured with respect to the same intrinsic nonlinearities that turn coherent states into something else. The experimental challenge here is confronted at the forefront of a number of technologies, involving the design of electromagnetic cavities, the cooling and trapping of atoms, and the development of microscopic lasers and novel optical materials. The theoretical challenge is that the conventional way of analyzing and thinking about open systems in quantum optics, based on the connection with a classical diffusion process, is now completely inadequate. These lectures are intended to state the challenge and describe some ideas that begin to met it.

The ultimate goal of the lectures is to discuss a new way of analyzing and thinking about the master equations that describe sources of light. The new approach employs what I call *quantum trajectories.* The words "quantum trajectory" refer to the path of a stochastic wavefunction that describes the state of an optical source, conditioned at each instant on a history of classical stochastic signals realized at ideal detectors monitoring the fields radiated by the source. Before we get to the new ideas, however, we must learn something about the master equations themselves and the standard ways in which they are analyzed. The first four lectures will deal with these subjects. The next two lectures deal with the theory of photoelectric detection, which we will use as a bridge between the old and the new. The quantum trajectory idea will occupy us for the final four lectures.

Contents

Introduction

The theory of open systems has been a theme in quantum optics since the birth of the subject some thirty years ago. The principal reason for this is that quantum optics was formed as a discipline around the invention of a new source of light – the laser. Sources of light are open systems. Thus, those working on the quantum theory of the laser found that they needed a way to treat dissipation in a quantum mechanical way [1]. The central ideas of a dissipative process are embodied in Fermi's golden rule and were used in quantum mechanics for many years before the invention of the laser. A complete theory of the laser needed more than this, however. Fermi's golden rule gives us a picture of quantum dynamics expressed in terms of transitions between discrete states. Needless to say, the fundamental equations of quantum mechanics are not expressed in these terms; they describe the continuous evolution of complex amplitudes, the coefficients in a superposition of states. With its emphasis on coherence, laser dynamics involves the complex amplitudes; to understand the properties of laser light something more than a calculation of rates for the incoherent emission and absorption of photons is required; the quantum theory of the laser must be formulated in a way that reveals the roles of both coherence and incoherence in open system dynamics.

Over the years, research in quantum optics has continued to be concerned with new sources of light. The current interest in squeezed light is, perhaps, the most prominent example. In this case the sources are parametric amplifiers, parametric oscillators, multi-wave mixers, and semiconductor diode lasers. Other fashionable topics in quantum optics, while not dealing directly with the development of practical light sources, have, nonetheless, been concerned with sources of light: resonance fluorescence, optical bistability, superradiance and superfluorescence are examples. Still further removed, sometimes it is the response of a system – an atom, or collection of atoms – to illumination by a source of light that is of interest, rather than the properties of the source itself. Problems of this sort also lend themselves to an open systems treatment. Collecting all the examples together, it might be said that the majority of calculations in quantum optics have a natural formulation in open systems language; those that are excluded are chiefly problems conceived in a rather idealized way in terms of a one- or few-mode interaction taking place in a lossless cavity.

Calculations in quantum optics use a wide variety of theoretical methods. They might solve a Schrödinger equation, a set of Heisenberg equations, or evaluate terms in a perturbation expansion; they might draw on techniques from atomic physics or on aids, such as Feynman diagrams, from quantum field theory. When, however, we look to the theme of open systems there are two principal methods that define something like a consistent language for the subject. The first is the quantum Langevin equation method, based on the Heisenberg equations of motion, and the second is the master equation method, rooted in the Schrödinger or wavefunction picture. Of course, the two approaches are closely related, and a preference to emphasize one or the other is determined to a large degree by personal taste. A more object choice can sometimes be made on the basis of ease of calculation, since there are nonlinear problems where the operator Langevin equations cannot be solved while it is possible to cast the master equation into a form that can be solved analytically. These lectures are about the master equation treatment of open systems in quantum optics. Some attention is also paid to Heisenberg equations, however, since they are used to define the multi-time averages, or correlation functions, of the fields radiated by a source.

When we speak of master equations in quantum optics the laser, once again, appears at the start of the history. As an exercise in quantum field theory the problem of laser light introduced one obvious novelty in comparison with more traditional applications of Q.E.D.: The field inside a laser cavity involves a very large number of photons; traditional Q.E.D. dealt with perturbative calculations and a few photons at most. Glauber developed the language used to describe laser light in terms of the highly excited states of a quantum field, building on the idea of the coherent state as the closest analogue of a classical field of fixed amplitude and phase [2-4]. The principal methods used to analyze master equations in quantum optics over the last thirty years are derivatives of Glauber's coherent state language. They are known generically as phase-space methods and are useful because they allow an operator master equation to be rewritten to look like a classical Fokker-Planck equation. The most widely known example of these methods is based on the diagonal coherent state representation proposed by Glauber [2, 4] and Sudarshan [5]. There are many generalizations of the approach. In all cases the utility of the approach can be traced to the same supposition – that at some acceptable level of approximation the phase-space representation allows the quantum-mechanical master equation to be replaced by a Fokker-Planck equation describing a classical diffusion process.

It was clear from the outset that most quantum states do not possess the simple form of coherent state representation that underlies the dynamical picture based on classical diffusion [6-9]. Glauber's P distribution was invented to represent mixtures of coherent states; in the world created by lasers it found many applications since conventional lasers do radiate mixtures of coherent states. Nevertheless, nonlinear interactions convert laser light into something other than a mixture of coherent states. In situations

Lecture 1 – Master Equations and Sources I

1.1 Photoemissive sources

Most experiments in quantum optics involve sources of light that might be called photoemissive sources. These are sources that emit photons irreversibly, to propagate away from the source until they are absorbed in the walls of the laboratory or are detected. Contrast this with the idealized picture of an electromagnetic field confined within a perfect cavity and measured inside the cavity by a detector introduced into that otherwise lossless environment. In the first scenario the act of detecting photons does not directly interfere with the source, since the photons have already left the source, irreversibly, before they encounter the detector. In the second, the detector is a major intrusion; if it turns photons into photoelectrons it removes photons from an otherwise lossless cavity; the description of the cavity field dynamics when the detector is present must be quite different from the description when it is not.

We are going to be discussing photoemissive sources and the quantum statistical methods that are used to analyze the properties of the light emitted by these sources. These sources are open systems that can be treated by the methods used to deal with dissipation in quantum mechanics. These methods allow us to make a convenient separation between the treatment of the source, as a damped quantum system, and the treatment of the fields the source emits. We begin by separating the whole system of source plus emitted fields into a system S, which accounts for all the internal workings of the source, and a reservoir R that carries the emitted fields; in the language of dissipative systems the reservoir describes the environment into which the source losses energy. For photoemissive sources the lost energy is useful energy – it appears in the reservoir as emitted fields that can be observed by an experimenter, or redirected for use elsewhere.

In this first lecture we will see a little of both ends of this problem. First we will discuss the master equation approach for treating a system (source) S damped by a reservoir R. We will then consider the problem of constructing the emitted fields in terms of the variables of the system S. In the first part of the lecture we follow the ideas for treating dissipation in quantum mechanics pioneered by Senitzky [1.1]. Other useful references are standard texts such as those by Louisell [1.2], and Sargent, Scully and Lamb [1.3].

1.2 Master equations

We begin with a Hamiltonian in the general form

$$H = H_S + H_R + H_{SR}, \tag{1.1}$$

where H_S and H_R are Hamiltonians for S and R, respectively, and H_{SR} is an interaction Hamiltonian. We will let $\chi(t)$ be the density operator for $S \oplus R$ and define the reduced density operator $\rho(t)$ by

$$\rho(t) = \mathrm{tr}_R[\chi(t)], \tag{1.2}$$

where the trace is only taken over the reservoir states. Clearly, if $\hat{O}$ is an operator in the Hilbert space of S we can calculate its average in the Schrödinger picture if we have knowledge of $\rho(t)$ alone, and not of the full $\chi(t)$:

$$\langle \hat{O} \rangle = \mathrm{tr}_{S\oplus R}[\hat{O}\chi(t)] = \mathrm{tr}_S\{\hat{O}\mathrm{tr}_R[\chi(t)]\} = \mathrm{tr}_S[\hat{O}\rho(t)]. \tag{1.3}$$

Our objective is to obtain an equation for $\rho(t)$ with the properties of R entering only as parameters.

The Schrödinger equation for χ reads

$$\dot{\chi} = \frac{1}{i\hbar}[H, \chi], \tag{1.4}$$

where H is given by (1.1). We transform (1.4) into the interaction picture, separating the rapid motion generated by $H_S + H_R$ from the slow motion generated by the interaction H_{SR}. With

$$\tilde{\chi}(t) = e^{(i/\hbar)(H_S+H_R)t}\chi(t)e^{-(i/\hbar)(H_S+H_R)t}, \tag{1.5}$$

from (1.1) and (1.4) we obtain

$$\dot{\tilde{\chi}} = \frac{1}{i\hbar}[\tilde{H}_{SR}(t), \tilde{\chi}], \tag{1.6}$$

where $\tilde{H}_{SR}(t)$ is explicitly time-dependent:

$$\tilde{H}_{SR}(t) = e^{(i/\hbar)(H_S+H_R)t}H_{SR}e^{-(i/\hbar)(H_S+H_R)t}. \tag{1.7}$$

We now integrate (1.6) formally to give

$$\tilde{\chi}(t) = \chi(0) + \frac{1}{i\hbar}\int_0^t dt' \, [\tilde{H}_{SR}(t'), \tilde{\chi}(t')], \tag{1.8}$$

and substitute for $\tilde{\chi}(t)$ inside the commutator in (1.6):

$$\dot{\tilde{\chi}} = \frac{1}{i\hbar}[\tilde{H}_{SR}(t), \chi(0)] - \frac{1}{\hbar^2}\int_0^t dt' \, [\tilde{H}_{SR}(t), [\tilde{H}_{SR}(t'), \tilde{\chi}(t')]]. \tag{1.9}$$

This equation is exact. Equation (1.4) has simply been cast into a convenient form where we can now identify reasonable approximations.

We will assume that the interaction is turned on at $t = 0$ and that no correlations exist between S and R at this initial time. Then $\chi(0) = \tilde{\chi}(0)$ factorizes as

$$\chi(0) = \rho(0)R_0, \tag{1.10}$$

where R_0 is an initial reservoir density operator. Then, noting that

$$\mathrm{tr}_R(\tilde{\chi}) = e^{(i/\hbar)H_S t}\rho e^{-(i/\hbar)H_S t} = \tilde{\rho}, \tag{1.11}$$

after tracing over the reservoir, (1.9) gives

$$\dot{\tilde{\rho}} = -\frac{1}{\hbar^2}\int_0^t dt' \,\mathrm{tr}_R\{[\tilde{H}_{SR}(t),[\tilde{H}_{SR}(t'),\tilde{\chi}(t')]]\}, \tag{1.12}$$

where, for simplicity, we have eliminated the term $(1/i\hbar)\mathrm{tr}_R\{[\tilde{H}_{SR}(t),\chi(0)]\}$ with the assumption $\mathrm{tr}_R[\tilde{H}_{SR}(t)R_0] = 0$. This is guaranteed if the reservoir operators coupling to S have zero mean in the state R_0, a condition which can always be arranged by simply including $\mathrm{tr}_R(H_{SR}R_0)$ in the system Hamiltonian.

We have stated that $\tilde{\chi}$ factorizes at $t = 0$. At later times correlations between S and R may arise due to the coupling of the system and reservoir through H_{SR}. However, we have assumed that this coupling is very weak, and at all times $\chi(t)$ should only show deviations of order H_{SR} from an uncorrelated state. Furthermore, R is a large system whose state should be virtually unaffected by its coupling to S. We then write

$$\tilde{\chi}(t) = \tilde{\rho}(t)R_0 + O(H_{SR}). \tag{1.13}$$

We now make our first major approximation, a *Born approximation*. Neglecting terms higher than second order in H_{SR}, we write (1.12) as

$$\dot{\tilde{\rho}} = -\frac{1}{\hbar^2}\int_0^t dt' \,\mathrm{tr}_R\{[\tilde{H}_{SR}(t),[\tilde{H}_{SR}(t'),\tilde{\rho}(t')R_0]]\}. \tag{1.14}$$

A detailed discussion of this approximation can be found in the work of Haake [1.4, 1.5].

Equation (1.14) is still a complicated equation. In particular, it is not Markoffian since the future evolution of $\tilde{\rho}(t)$ depends on its past history through the integration over $\tilde{\rho}(t')$ (the future behavior of a Markoffian system depends only on its present state). Our second major approximation, the *Markoff approximation*, replaces $\tilde{\rho}(t')$ by $\tilde{\rho}(t)$ to obtain a *master equation in the Born-Markoff approximation*:

$$\dot{\tilde{\rho}} = -\frac{1}{\hbar^2}\int_0^t dt' \,\mathrm{tr}_R\{[\tilde{H}_{SR}(t),[\tilde{H}_{SR}(t'),\tilde{\rho}(t)R_0]]\}. \tag{1.15}$$

Markoffian behavior seems reasonable on physical grounds. Potentially, S can depend on its past history because its earlier states become imprinted as changes in the reservoir state through the interaction H_{SR}; earlier states are then reflected back on the future evolution of S as it interacts with the changed reservoir. If, however, the reservoir is a large system maintained in thermal equilibrium, we do not expect it to preserve the minor changes brought by its interaction with S for very long; not for long enough to significantly affect the future evolution of S. It becomes a question of reservoir correlation time versus the time scale for significant change in S. By studying the integrand of (1.14) with this view in mind we can make the underlying assumption of the Markoff approximation more explicit.

We first make our model a little more specific by writing

$$H_{SR} = \hbar \sum_i s_i \Gamma_i, \tag{1.16}$$

where the s_i are operators in the Hilbert space of S and the Γ_i are operators in the Hilbert space of R. Then

$$\tilde{H}_{SR}(t) = \hbar \sum_i \tilde{s}_i(t)\tilde{\Gamma}_i(t). \tag{1.17}$$

The master equation in the Born approximation is now

$$\begin{aligned}\dot{\tilde{\rho}} &= -\sum_{i,j} \int_0^t dt' \operatorname{tr}_R\{[\tilde{s}_i(t)\tilde{\Gamma}_i(t), [\tilde{s}_j(t')\tilde{\Gamma}_j(t'), \tilde{\rho}(t')R_0]]\} \\ &= -\sum_{i,j} \int_0^t dt' \Big\{[\tilde{s}_i(t)\tilde{s}_j(t')\tilde{\rho}(t') - \tilde{s}_j(t')\tilde{\rho}(t')\tilde{s}_i(t)]\langle\tilde{\Gamma}_i(t)\tilde{\Gamma}_j(t')\rangle_R \\ &\quad + [\tilde{\rho}(t')\tilde{s}_j(t')\tilde{s}_i(t) - \tilde{s}_i(t)\tilde{\rho}(t')\tilde{s}_j(t')]\langle\tilde{\Gamma}_j(t')\tilde{\Gamma}_i(t)\rangle_R\Big\},\end{aligned} \tag{1.18}$$

where we have used the cyclic property of the trace – $\operatorname{tr}(ABC) = \operatorname{tr}(CAB) = \operatorname{tr}(BCA)$ – and write

$$\langle\tilde{\Gamma}_i(t)\tilde{\Gamma}_j(t')\rangle_R = \operatorname{tr}_R[R_0\tilde{\Gamma}_i(t)\tilde{\Gamma}_j(t')], \tag{1.19a}$$

$$\langle\tilde{\Gamma}_j(t')\tilde{\Gamma}_i(t)\rangle_R = \operatorname{tr}_R[R_0\tilde{\Gamma}_j(t')\tilde{\Gamma}_i(t)]. \tag{1.19b}$$

The properties of the reservoir enter (1.18) through the two correlation functions (1.19a) and (1.19b). We can justify the replacement of $\tilde{\rho}(t')$ by $\tilde{\rho}(t)$ if these correlation functions decay very rapidly on the time scale on which $\tilde{\rho}(t)$ varies. Ideally, we might take

$$\langle\tilde{\Gamma}_i(t)\tilde{\Gamma}_j(t')\rangle_R \propto \delta(t-t'). \tag{1.20}$$

The Markoff approximation then relies, as suggested, on the existence of two widely separated time scales: a slow time scale for the dynamics of the system S, and a fast time scale characterizing the decay of reservoir correlation functions. Further discussion on this point is given by Schieve and Middleton [1.6].

1.3 Master equation for a cavity mode driven by thermal light

Let us consider an explicit model. We will derive the mater equation for a single mode of the optical cavity illustrated in Fig. 1.1. The figure shows a ring cavity with the reservoir comprised of travelling-wave modes that satisfy periodic boundary conditions at $z = -L'/2$ and $z = L'/2$. The cavity mode (system S) couples to the reservoir through a partially transmitting mirror. For the Hamiltonian of the composite system $S \oplus R$ we write

$$H_S = \hbar\omega_C a^\dagger a, \tag{1.21a}$$

$$H_R = \sum_j \hbar\omega_j {r_j}^\dagger r_j, \tag{1.21b}$$

$$H_{SR} = \sum_j \hbar(\kappa_j^* a {r_j}^\dagger + \kappa_j a^\dagger r_j) = \hbar(a\Gamma^\dagger + a^\dagger \Gamma). \tag{1.21c}$$

The system S is an harmonic oscillator with frequency ω_C and creation and annihilation operators $a^\dagger$ and a, respectively; the reservoir R is a collection of harmonic oscillators with frequencies ω_j, and corresponding creation and annihilation operators ${r_j}^\dagger$ and r_j, respectively; the oscillator a couples to the jth reservoir oscillator via a coupling constant κ_j (for the moment unspecified) in the rotating-wave approximation. We take the reservoir to be in thermal equilibrium at temperature T – the cavity mode is driven by thermal light:

$$R_0 = \prod_j e^{-\hbar\omega_j {r_j}^\dagger r_j / k_B T}\left(1 - e^{-\hbar\omega_j / k_B T}\right), \tag{1.22}$$

where k_B is Boltzmann's constant.

The identification with (1.18) is made by setting

$$s_1 = a, \qquad s_2 = a^\dagger, \tag{1.23a}$$

$$\Gamma_1 = \Gamma^\dagger = \sum_j \kappa_j^* {r_j}^\dagger, \qquad \Gamma_2 = \Gamma = \sum_j \kappa_j r_j, \tag{1.23b}$$

and the operators in the interaction picture are

$$\tilde{s}_1(t) = e^{i\omega_C a^\dagger a t} a e^{-i\omega_C a^\dagger a t} = a e^{-i\omega_C t}, \tag{1.24a}$$

$$\tilde{s}_2(t) = e^{i\omega_C a^\dagger a t} a^\dagger e^{-i\omega_C a^\dagger a t} = a^\dagger e^{i\omega_C t}, \tag{1.24b}$$

and

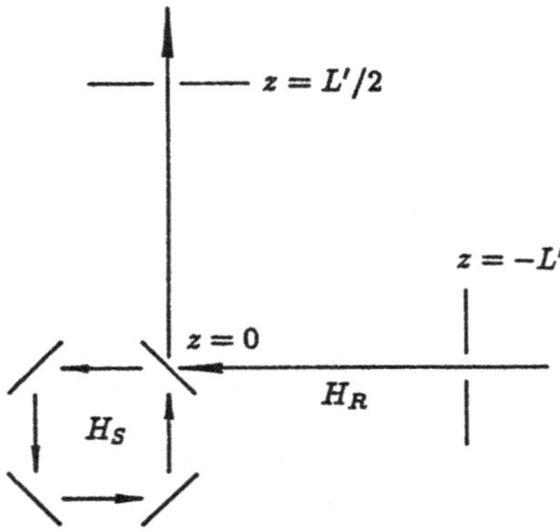

Fig. 1.1. Schematic diagram of a cavity mode coupled to a travelling-wave reservoir

$$\tilde{\Gamma}_1(t) = \tilde{\Gamma}^\dagger(t) = \exp\left(i\sum_n \omega_n r_n{}^\dagger r_n t\right)\sum_j \kappa_j^* r_j{}^\dagger \exp\left(-i\sum_m \omega_m r_m{}^\dagger r_m t\right)$$

$$= \sum_j \kappa_j^* r_j{}^\dagger e^{i\omega_j t}, \tag{1.25a}$$

$$\tilde{\Gamma}_2(t) = \tilde{\Gamma}(t) = \sum_j \kappa_j r_j e^{-i\omega_j t}. \tag{1.25b}$$

Now, since the summation in (1.18) runs over $i = 1,2$ and $j = 1,2$, the integrand involves sixteen terms. We write

$$\begin{aligned}\dot{\tilde{\rho}} = -\int_0^t dt' \Big\{ &\left[aa\tilde{\rho}(t') - a\tilde{\rho}(t')a\right] e^{-i\omega_C(t+t')}\langle\tilde{\Gamma}^\dagger(t)\tilde{\Gamma}^\dagger(t')\rangle_R + \text{h.c.} \\ &+ \left[a^\dagger a^\dagger\tilde{\rho}(t') - a^\dagger\tilde{\rho}(t')a^\dagger\right] e^{i\omega_C(t+t')}\langle\tilde{\Gamma}(t)\tilde{\Gamma}(t')\rangle_R + \text{h.c.} \\ &+ \left[aa^\dagger\tilde{\rho}(t') - a^\dagger\tilde{\rho}(t')a\right] e^{-i\omega_C(t-t')}\langle\tilde{\Gamma}^\dagger(t)\tilde{\Gamma}(t')\rangle_R + \text{h.c.} \\ &+ \left[a^\dagger a\tilde{\rho}(t') - a\tilde{\rho}(t')a^\dagger\right] e^{i\omega_C(t-t')}\langle\tilde{\Gamma}(t)\tilde{\Gamma}^\dagger(t')\rangle_R + \text{h.c.}\Big\},\end{aligned} \tag{1.26}$$

where the reservoir correlation functions are explicitly:

$$\langle\tilde{\Gamma}^\dagger(t)\tilde{\Gamma}^\dagger(t')\rangle_R = 0, \tag{1.27}$$

$$\langle\tilde{\Gamma}(t)\tilde{\Gamma}(t')\rangle_R = 0, \tag{1.28}$$

$$\langle\tilde{\Gamma}^\dagger(t)\tilde{\Gamma}(t')\rangle_R = \sum_j |\kappa_j|^2 e^{i\omega_j(t-t')}\bar{n}(\omega_j, T), \tag{1.29}$$

$$\langle\tilde{\Gamma}(t)\tilde{\Gamma}^\dagger(t')\rangle_R = \sum_j |\kappa_j|^2 e^{-i\omega_j(t-t')}[\bar{n}(\omega_j, T) + 1], \tag{1.30}$$

with

$$\bar{n}(\omega_j, T) = \text{tr}_R(R_0 r_j{}^\dagger r_j) = \frac{e^{-\hbar\omega_j/k_B T}}{1 - e^{-\hbar\omega_j/k_B T}}. \tag{1.31}$$

These results are obtained quite readily if the trace is taken using the multimode Fock states. $\bar{n}(\omega_j, T)$ is the mean photon number for an oscillator with frequency ω_j in thermal equilibrium at temperature T.

The nonvanishing reservoir correlation functions (1.29) and (1.30) involve a summation over the reservoir oscillators. We will change this summation to an integration by introducing a density of states $g(\omega)$, such that $g(\omega)d\omega$ gives the number of oscillators with frequencies in the interval ω to $\omega + d\omega$. For the one-dimensional reservoir field illustrated in Fig. 1.1,

$$g(\omega) = L'/2\pi c. \tag{1.32}$$

Making the change of variable $\tau = t - t'$ in (1.26), this equation can then be restated as

$$\begin{aligned}\dot{\tilde{\rho}} = -\int_0^t d\tau \Big\{ \Big[a a^\dagger \tilde{\rho}(t-\tau) - a^\dagger \tilde{\rho}(t-\tau) a \Big] e^{-i\omega_C \tau} \langle \tilde{\Gamma}^\dagger(t) \tilde{\Gamma}(t-\tau) \rangle_R + \text{h.c.} \\ + \Big[a^\dagger a \tilde{\rho}(t-\tau) - a \tilde{\rho}(t-\tau) a^\dagger \Big] e^{i\omega_C \tau} \langle \tilde{\Gamma}(t) \tilde{\Gamma}^\dagger(t-\tau) \rangle_R + \text{h.c.} \Big\}, \end{aligned} \tag{1.33}$$

where the nonzero reservoir correlation functions are

$$\langle \tilde{\Gamma}^\dagger(t) \tilde{\Gamma}(t-\tau) \rangle_R = \int_0^\infty d\omega\, e^{i\omega\tau} g(\omega) |\kappa(\omega)|^2 \bar{n}(\omega, T), \tag{1.34}$$

$$\langle \tilde{\Gamma}(t) \tilde{\Gamma}^\dagger(t-\tau) \rangle_R = \int_0^\infty d\omega\, e^{-i\omega\tau} g(\omega) |\kappa(\omega)|^2 [\bar{n}(\omega, T) + 1], \tag{1.35}$$

with $\bar{n}(\omega, T)$ given by (1.31) with ω_j replaced by ω.

We can now argue more specifically about the Markoff approximation. Are (1.34) and (1.35) approximately proportional to $\delta(\tau)$? We can certainly see that for τ "large enough" the oscillating exponential will average the "slowly varying" functions $g(\omega)$, $|\kappa(\omega)|^2$, and $\bar{n}(\omega, T)$ essentially to zero. However, how large is large enough? If we can argue for a specific ω dependence in $\kappa(\omega)$, then with (1.31) and (1.32) we can evaluate the correlation functions (1.34) and (1.35) explicitly to obtain the reservoir correlation time. The Markoff approximation assumes that this correlation time is very short compared to the time scale for significant change in $\tilde{\rho}$. Since we do not yet have an equation of motion for $\tilde{\rho}$ we must rely on intuition rather than a solution for $\tilde{\rho}(t)$ to tell us how fast $\tilde{\rho}$ will change. The free oscillation at the frequency ω_C is removed by the transformation to the interaction picture; therefore, we expect that the remaining time dependence in $\tilde{\rho}$ is characterized by a decay time for the cavity mode – by the inverse of the mode linewidth. This might typically be a number of the order of 10^{-8} sec.

To estimate the reservoir correlation time let us just take $\kappa(\omega)$ to be constant and focus on the frequency dependence of $\bar{n}(\omega, T)$. Because of the $e^{\pm i\omega_C \tau}$ multiplying the reservoir correlation functions in (1.33) it is really only the $\omega \approx \omega_C$ part of the frequency range in (1.34) and (1.35) that is important. We can therefore estimate the reservoir correlation time by extending the frequency integrals to $-\infty$ [replace $\bar{n}(\omega, T)$ by $\bar{n}(|\omega|, T)$]. We now have a Fourier transform and the correlation time will be given by the inverse width $\hbar/k_BT$ of the function $\bar{n}(|\omega|,T)$. At room temperature this gives a number of the order of 0.25×10^{-13} sec, much less than our estimated time scale for significant changes to occur in $\tilde{\rho}$. [Under the assumption that $\kappa(\omega)$ is constant the $+1$ in (1.35) adds a δ-function.]

Now we are satisfied that the τ integration in (1.33) is dominated by times that are much shorter than the time scale for the evolution of $\tilde{\rho}$, we replace $\tilde{\rho}(t-\tau)$ by $\tilde{\rho}(t)$ and obtain

$$\dot{\tilde{\rho}} = \alpha(a\tilde{\rho}a^\dagger - a^\dagger a\tilde{\rho}) + \beta(a\tilde{\rho}a^\dagger + a^\dagger\tilde{\rho}a - a^\dagger a\tilde{\rho} - \tilde{\rho}aa^\dagger) + \text{h.c.}, \tag{1.36}$$

with

$$\alpha = \int_0^t d\tau \int_0^\infty d\omega\, e^{-i(\omega-\omega_C)\tau} g(\omega)|\kappa(\omega)|^2, \tag{1.37}$$

$$\beta = \int_0^t d\tau \int_0^\infty d\omega\, e^{-i(\omega-\omega_C)\tau} g(\omega)|\kappa(\omega)|^2 \bar{n}(\omega,T). \tag{1.38}$$

Then, since t is a time typical of the time scale for changes in $\tilde{\rho}$, and the τ integration is dominated by much shorter times characterizing the decay of reservoir correlations, we can extend the τ integration to infinity and evaluate α and β using

$$\lim_{t\to\infty} \int_0^t d\tau\, e^{-i(\omega-\omega_C)\tau} = \pi\delta(\omega-\omega_C) + i\frac{P}{\omega_C-\omega}, \tag{1.39}$$

where P indicates the Cauchy principal value. We find

$$\alpha = \pi g(\omega_C)|\kappa(\omega_C)|^2 + i\Delta, \tag{1.40}$$

$$\beta = \pi g(\omega_C)|\kappa(\omega_C)|^2 \bar{n}(\omega_C) + i\Delta', \tag{1.41}$$

with

$$\Delta = P\int_0^\infty d\omega\, \frac{g(\omega)|\kappa(\omega)|^2}{\omega_C-\omega}, \tag{1.42}$$

$$\Delta' = P\int_0^\infty d\omega\, \frac{g(\omega)|\kappa(\omega)|^2}{\omega_C-\omega}\bar{n}(\omega,T). \tag{1.43}$$

We finally have our master equation. Setting

$$\kappa = \pi g(\omega_C)|\kappa(\omega_C)|^2, \qquad \bar{n} = \bar{n}(\omega_C,T), \tag{1.44}$$

from (1.36), (1.40), and (1.41), we obtain

$$\begin{aligned}\dot{\tilde{\rho}} =& -i\Delta[a^\dagger a,\tilde{\rho}] + \kappa(2a\tilde{\rho}a^\dagger - a^\dagger a\tilde{\rho} - \tilde{\rho}a^\dagger a)\\ &+ 2\kappa\bar{n}(a\tilde{\rho}a^\dagger + a^\dagger\tilde{\rho}a - a^\dagger a\tilde{\rho} - \tilde{\rho}aa^\dagger).\end{aligned} \tag{1.45}$$

Here $\tilde{\rho}$ is still in the interaction picture. To transform back to the Schrödinger picture (1.11) gives

$$\dot{\rho} = \frac{1}{i\hbar}[H_S, \rho] + e^{-(i/\hbar)H_S t}\dot{\tilde{\rho}}e^{(i/\hbar)H_S t}. \tag{1.46}$$

With $H_S = \hbar\omega_C a^\dagger a$, we substitute for $\dot{\tilde{\rho}}$ and use (1.11) and (1.24) to arrive at the *master equation for a cavity mode driven by thermal light*:

$$\begin{aligned}\dot{\rho} = &- i\omega_C'[a^\dagger a, \rho] + \kappa(2a\rho a^\dagger - a^\dagger a\rho - \rho a^\dagger a)\\ &+ 2\kappa\bar{n}(a\rho a^\dagger + a^\dagger\rho a - a^\dagger a\rho - \rho a a^\dagger),\end{aligned} \tag{1.47}$$

where

$$\omega_C' = \omega_C + \Delta. \tag{1.48}$$

1.4 The cavity output field

The master equation (1.47) provides a description of the field inside the (lossy) cavity. Normally we would want to observe the field from outside the cavity. The cavity mode is a source, radiating a field that is carried by the modes of the reservoir. Classically, the field at the output of an optical cavity is obtained from the intracavity field after multiplying by a mirror transmission coefficient. Quantum mechanically this simple relationship will not do. It asserts that the output field is described by operators $\sqrt{T}e^{i\phi_T}a$ and $\sqrt{T}e^{-i\phi_T}a^\dagger$, where T is the transmission coefficient of the output mirror and ϕ_T is the phase change on transmission through the output mirror. But a and $a^\dagger$ obey the commutation relation $[a, a^\dagger] = 1$, and therefore $[\sqrt{T}e^{i\phi_T}a, \sqrt{T}e^{-i\phi_T}a^\dagger] = T < 1$. Thus, special care must be taken to preserve commutation relations.

We can construct the cavity output field by calculating the source contribution – the contribution from S – to the reservoir mode operators r_k and $r_k^\dagger$. The field outside the cavity is described by the Heisenberg operator

$$\hat{\boldsymbol{E}}(z,t) = \hat{\boldsymbol{E}}^{(+)}(z,t) + \hat{\boldsymbol{E}}^{(-)}(z,t), \tag{1.49a}$$

with

$$\hat{\boldsymbol{E}}^{(+)}(z,t) = i\hat{e}_0 \sum_k \sqrt{\frac{\hbar\omega_k}{2\epsilon_0 AL'}}\, r_k(t) e^{i[(\omega_k/c)z + \phi(z)]}, \tag{1.49b}$$

$$\hat{\boldsymbol{E}}^{(-)}(z,t) = \hat{\boldsymbol{E}}^{(+)}(z,t)^\dagger, \tag{1.49c}$$

where

$$\phi(z) = \begin{cases}\phi_R & z > 0\\ 0 & z < 0\end{cases}; \tag{1.50}$$

ϕ_R is the phase change on reflection at the cavity output mirror and A is the cross-sectional area of the cavity mode. Using the Hamiltonian (1.21), we obtain Heisenberg equations of motion

$$\dot{r}_k = -i\omega_k r_k - i\kappa_k^* a. \tag{1.51}$$

The term $i\kappa_k^* a$ couples energy from the intracavity field into the modes of the external field; for the present the coupling constant κ_k^* is left unspecified. Integrating (1.51) formally, we have

$$r_k(t) = r_k(0)e^{-i\omega_k t} - i\kappa_k^* e^{-i\omega_C t}\int_0^t dt'\,\tilde{a}(t')e^{i(\omega_k-\omega_C)(t'-t)}, \tag{1.52}$$

where $\tilde{a}(t)$ is the slowly-varying operator

$$\tilde{a}(t) = e^{i\omega_C t}a(t). \tag{1.53}$$

Then the laser output field is given by

$$\hat{\boldsymbol{E}}^{(+)}(z,t) = \hat{\boldsymbol{E}}_f^{(+)}(z,t) + \hat{\boldsymbol{E}}_s^{(+)}(z,t), \tag{1.54}$$

with

$$\hat{\boldsymbol{E}}_f^{(+)}(z,t) = i\hat{e}_0\sum_k\sqrt{\frac{\hbar\omega_k}{2\epsilon_0 AL'}}\,r_k(0)e^{-i[\omega_k(t-z/c)-\phi(z)]}, \tag{1.55}$$

and

$$\begin{aligned}\hat{\boldsymbol{E}}_s^{(+)}(z,t) = {}& \hat{e}_0\sqrt{\frac{\hbar}{2\epsilon_0 AL'}}\,e^{-i[\omega_C(t-z/c)-\phi(z)]}\\ &\times\sum_k\sqrt{\omega_k}\,\kappa_k^*\int_0^t dt'\,\tilde{a}(t')e^{i(\omega_k-\omega_C)(t'-t+z/c)}.\end{aligned} \tag{1.56}$$

This field decomposes into the sum of a freely evolving field $\hat{\boldsymbol{E}}_f^{(+)}(z,t)$, and a source field $\hat{\boldsymbol{E}}_s^{(+)}(z,t)$.

To express the source field in manageable form we introduce the mode density (1.32) and perform the summation over k as an integral:

$$\begin{aligned}\hat{\boldsymbol{E}}_s^{(+)}(z,t) = {}& \hat{e}_0\sqrt{\frac{\hbar}{2\epsilon_0 Ac}}\sqrt{\frac{L'}{c}}\,e^{-i[\omega_C(t-z/c)-\phi(z)]}\\ &\times\frac{1}{2\pi}\int_0^\infty d\omega\,\sqrt{\omega}\,\kappa^*(\omega)\int_0^t dt'\,\tilde{a}(t')e^{i(\omega-\omega_C)(t'-t+z/c)}.\end{aligned} \tag{1.57}$$

Now, since we have removed the rapid oscillation at the cavity resonance frequency in (1.53), $\tilde{a}$ is expected to vary slowly in comparison with the optical period – on a time scale characterized by κ^{-1} [see the discussion below (1.35)]. Thus, for frequencies outside the range $-100\kappa \le \omega - \omega_C \le 100\kappa$, say, the time integral in (1.57) averages to zero. This means that over the important range of the frequency integral we can assume that $\sqrt{\omega}\kappa^*(\omega) \approx \sqrt{\omega_C}\kappa^*(\omega_C)$; we can also extend the range of the frequency integral to $-\infty$. Then, evaluating the frequency integral, we obtain

$$
\begin{aligned}
\hat{E}_s^{(+)}(z,t) \\
&= \hat{e}_0\sqrt{\frac{\hbar\omega_C}{2\epsilon_0 Ac}}\sqrt{\frac{L'}{c}}\kappa^*(\omega_C)e^{-i[\omega_C(t-z/c)-\phi(z)]}\int_0^t dt'\,\tilde{a}(t')\delta(t'-t+z/c) \\
&= \begin{cases} \hat{e}_0\sqrt{\dfrac{\hbar\omega_C}{2\epsilon_0 Ac}}\sqrt{\dfrac{L'}{c}}\kappa^*(\omega_C)e^{i\phi_R}a(t-z/c) & ct>z>0 \\ 0 & z<0. \end{cases}
\end{aligned}
\tag{1.58}
$$

Thus, for $ct > z > 0$ the source field is proportional to the intracavity field evaluated at the retarded time $t - z/c$.

We can now fix the value of the reservoir coupling constant $\kappa^*(\omega_C)$. If (1.58) is to give the expected relationship $\langle a\rangle \to \sqrt{T}e^{i\phi_T}\langle a\rangle$ between the mean intracavity field and the mean output field, we must choose

$$
-i\sqrt{\frac{L'}{c}}\kappa^*(\omega_C)e^{i\phi_R} = \sqrt{\frac{c}{L}}\sqrt{T}e^{i\phi_T} = \sqrt{2\kappa}\,e^{i\phi_T}, \tag{1.59}
$$

where $\kappa = Tc/2L$ is the cavity decay rate appearing in the master equation (1.47); L is the round-trip distance in the cavity. We can also derive this relationship (without the phase factor) from (1.44), which gives $2\kappa = 2\pi g(\omega_C)|\kappa(\omega_C)|^2$. Substituting (1.32) for the reservoir density of states, we find $\sqrt{L'/c}|\kappa(\omega_C)| = \sqrt{2\kappa}$, which is the modulus of the relationship (1.59). The final form for the source term in the cavity output field is now

$$
\hat{E}_s^{(+)}(z,t) = \begin{cases} i\hat{e}_0\sqrt{\dfrac{\hbar\omega_C}{2\epsilon_0 Ac}}\sqrt{2\kappa}\,e^{i\phi_T}a(t-z/c) & ct>z>0 \\ 0 & z<0. \end{cases} \tag{1.60}
$$

Equation (1.60) yields exactly what we would expect for the average photon flux from the cavity:

$$
\frac{2\epsilon_0 cA}{\hbar\omega_C}\langle\hat{E}_s^{(-)}(z,t)\hat{E}_s^{(+)}(z,t)\rangle = 2\kappa\langle a^\dagger(t-z/c)a(t-z/c)\rangle. \tag{1.61}
$$

The right-hand side is the product of the photon escape probability per unit time and the mean number of photons in the cavity.

In fact (1.60) is the relationship we would write down from the classical result for the transmission of the intracavity field through the cavity output mirror; we could have constructed the full expression (1.54) for the cavity output field from our understanding of the classical boundary conditions at the output mirror; the free-field term is just the contribution from the reflection of incoming reservoir modes into the cavity output (our theory assumes $R = 1-T \approx 1$). The only difference between the quantum-mechanical and classical pictures is that $\hat{E}_f^{(+)}(z,t)$ and $\hat{E}_s^{(+)}(z,t)$ are operators in the quantum-mechanical theory, and therefore play an algebraic role that is absent in a classical theory. The operators $\hat{E}_f^{(\pm)}(z,t)$ and $\hat{E}_s^{(\mp)}(z,t)$ do not commute; it is their noncommutation that preserves the commutation relation for the operators $\hat{\boldsymbol{E}}^{(+)}(z,t)$ and $\hat{\boldsymbol{E}}^{(-)}(z,t)$ of the total field. Thus, the free-field term cannot be dropped from (1.54) even when the reservoir modes are in the vacuum state. However, when the reservoir modes are in the vacuum state, this concern for algebraic integrity in the quantum theory really has little practical consequence, since we are generally interested in normal-ordered, time-ordered operator averages, quantities that are insensitive to vacuum contributions.

1.5 Correlations between the free field and the source field

When the free field (reservoir) is in the vacuum state and normal-ordered, time-ordered averages only are needed, connecting the statistical properties of the output field to the quantum dynamics of the source is a trivial exercise. We simply multiply the retarded intracavity field amplitude $i\hat{e}_0\sqrt{\hbar\omega_C/2\epsilon_0 AL}\,a(t-z/c)$ by $\sqrt{T}\,e^{i\phi_T} = \sqrt{L/c}\sqrt{2\kappa}\,e^{i\phi_T}$ to convert photon numbers inside the cavity to a photon flux outside, as in (1.60) and (1.61). But when the free field is not in the vacuum state, or non-normal-ordered or non-time-ordered averages are needed, things are not so straightforward. Then the free field contributes to the output, and to calculate its contribution we generally need nontrivial information about how it is correlated with the source.

Consider first an almost trivial example; consider an empty cavity driven by a coherent field. The reservoir mode with frequency $\omega_k = \omega_C$ is in the coherent state $|\beta\rangle$, and all other modes are in the vacuum state. Thus, from (1.54), (1.55), and (1.60), the cavity is driven on resonance by the mean field ($z < 0$)

$$\langle\hat{\boldsymbol{E}}^{(+)}(z,t)\rangle = \langle\hat{\boldsymbol{E}}_f^{(+)}(z,t)\rangle = i\hat{e}_0\sqrt{\frac{\hbar\omega_C}{2\epsilon_0 AL'}}\,\beta e^{-i\omega_C(t-z/c)}, \tag{1.62}$$

with mean output field ($z > 0$)

$$\begin{aligned}\langle\hat{\boldsymbol{E}}^{(+)}(z,t)\rangle &= \langle\hat{\boldsymbol{E}}_f^{(+)}(z,t)\rangle + \langle\hat{\boldsymbol{E}}_s^{(+)}(z,t)\rangle \\ &= i\hat{e}_0\sqrt{\frac{\hbar\omega_C}{2\epsilon_0 AL'}}\Big(e^{i\phi_R}\beta \\ &\quad +\sqrt{L'/c}\sqrt{2\kappa}\,e^{i\phi_T}\langle\tilde{a}(t-z/c)\rangle\Big)e^{-i\omega_C(t-z/c)}.\end{aligned} \tag{1.63}$$

The first term inside the bracket is the input field, reflected into the output, and the second term is the field radiated by the cavity. Since the cavity has only one partially transmitting mirror, in the steady state the two contributions must interfere to reconstruct the input amplitude, with a possible phase change. To check that this is so we need $\langle\tilde{a}\rangle_{ss}$. This is obtained from the mean-value equation

$$\langle\dot{\tilde{a}}\rangle = -\kappa\langle\tilde{a}\rangle - i\kappa(\omega_C)\beta, \tag{1.64}$$

where $\kappa(\omega_C)$ is the system-reservoir coupling coefficient given by (1.59); the driving term in (1.64) is derived from the interaction Hamiltonian $H_{SR}|_{\omega_k=\omega_C} = \hbar(\kappa_k^* a r_k^\dagger + \kappa_k a^\dagger r_k)|_{\omega_k=\omega_C}$. Substituting the steady-state solution $\langle\tilde{a}\rangle_{ss} = -i\kappa(\omega_C)\beta/\kappa$ to (1.64) into (1.63), we find $(z > 0)$

$$\begin{aligned}\langle\hat{\boldsymbol{E}}^{(+)}(z,t)\rangle &= i\hat{e}_0\sqrt{\frac{\hbar\omega_C}{2\epsilon_0 AL'}}\Bigg[e^{i\phi_R}\beta \\ &\qquad + \sqrt{L'/c}\sqrt{2\kappa}\,e^{i\phi_T}\left(-\sqrt{\frac{c}{L'}}\frac{\sqrt{2\kappa}\beta}{\kappa}e^{i(\phi_R-\phi_T)}\right)\Bigg]e^{-i\omega_C(t-z/c)} \\ &= -e^{i\phi_R}i\hat{e}_0\sqrt{\frac{\hbar\omega_C}{2\epsilon_0 AL'}}\beta e^{-i\omega_C(t-z/c)}.\end{aligned} \tag{1.65}$$

This is the mean driving field amplitude multiplied by the phase factor $-e^{i\phi_R}$. Thus, we do recover the anticipated result.

Accounting for free-field contributions is more difficult when this field is not in a coherent state. The master equation (1.49) was derived for a thermal reservoir, and reservoirs with different statistical properties are also sometimes of interest – for example, squeezed reservoirs, where the free field is in a broadband squeezed state. We can appreciate the difficulties that arise and the road to their resolution by considering the first-order correlation function for the full output field $\hat{\boldsymbol{E}}(z,t)$. First, let us simplify the notation in (1.54), (1.55), and (1.60) by scaling the field operators so that the source field appears in units of photon flux. We write $(ct > z > 0)$

$$\hat{\mathcal{E}}(z,t) = \sqrt{c/L'}\,r(t-z/c) + \sqrt{2\kappa}\,a(t-z/c), \tag{1.66}$$

where

$$\hat{\mathcal{E}}(z,t) = -ie^{-i\phi_T}\sqrt{\frac{2\epsilon_0 Ac}{\hbar\omega_C}}\,\hat{e}_0\cdot\hat{\boldsymbol{E}}^{(+)}(z,t), \tag{1.67a}$$

$$r_f(t-z/c) = e^{i(\phi_R-\phi_T)}\sum_k\sqrt{\frac{\omega_k}{\omega_C}}\,r_k(0)e^{-i\omega_k(t-z/c)}. \tag{1.67b}$$

Then the normalized first-order correlation function for the field $\hat{\boldsymbol{E}}(z,t)$ is given by

$$\begin{aligned}g_{ss}^{(1)}(\tau) = \left(\langle\hat{\mathcal{E}}^\dagger\hat{\mathcal{E}}\rangle_{ss}\right)^{-1}\Big\{&(c/L')\langle r_f^\dagger(0)r_f(\tau)\rangle + 2\kappa\Big[\lim_{t\to\infty}\langle a^\dagger(t)a(t+\tau)\rangle\Big] \\ &+ \sqrt{c/L'}\sqrt{2\kappa}\Big[\lim_{t\to\infty}\langle r_f^\dagger(t)a(t+\tau)\rangle + \lim_{t\to\infty}\langle a^\dagger(t)r_f(t+\tau)\rangle\Big]\Big\},\end{aligned} \tag{1.68}$$

with

$$\langle\hat{\mathcal{E}}^\dagger\hat{\mathcal{E}}\rangle_{ss} = (c/L')\langle r_f^\dagger r_f\rangle + 2\kappa\langle a^\dagger a\rangle_{ss} + \sqrt{c/L'}\sqrt{2\kappa}\Big(\langle r_f^\dagger a\rangle_{ss} + \langle a^\dagger r_f\rangle_{ss}\Big). \tag{1.69}$$

We need more than the source-field correlation function $\langle a^\dagger(t)a(t+\tau)\rangle$ if we are going to calculate this quantity. The free-field correlation function $\langle r_f^\dagger(t)r_f(t+\tau)\rangle$ is presumably straightforward to calculate, given the state of the reservoir. But how do we calculate the correlations between the free field and the source field, the correlation functions $\langle r_f^\dagger(t)a(t+\tau)\rangle$ and $\langle a^\dagger(t)r_f(t+\tau)\rangle$?

When the free field is in a coherent state these correlation functions factorize; because they are in normal order, action of $r_f^\dagger$ and r_f to the left and right, respectively, on the reservoir state, replaces the operators by coherent amplitudes. But in general there is no similarly straightforward procedure available. Gardiner and Collett [1.7] provide a method for calculating these correlation functions using an input-output theory built around quantum stochastic differential equations – a Heisenberg picture formulation of reservoir theory. A different approach that is more closely tied to the Schrödinger picture formulation of reservoir theory we are using is given by Carmichael [1.8]. We do not have time to go through the details of these calculations but can outline the basic idea.

We must begin the calculation at a level that still includes the reservoir operators explicitly. The master equation is of no direct use since the reservoir operators have been traced out of this equation. We return to the Heisenberg equations of motion. The Heisenberg equation for the mode operators of the reservoir field is given by (1.51). The Heisenberg equation for the cavity mode reads

$$\dot{a} = \frac{1}{i\hbar}[a, H_S] - i\sum_k \kappa_k r_k, \tag{1.70}$$

where we have used $H_{SR} = \hbar(a\Gamma^\dagger + a^\dagger\Gamma)$, with $\Gamma^\dagger$ and Γ given by (1.23b). Substituting the formal solution (1.52) for $r_k(t)$, and treating the mode summation and time integral as we did in passing from (1.57) to (1.58), we have

$$\begin{aligned}
\dot{a} &= \frac{1}{i\hbar}[a, H_S] - e^{-i\omega_C t}\sum_k |\kappa_k|^2 \int_0^t dt'\tilde{a}(t')e^{i(\omega_k-\omega_C)(t'-t)} \\
&\quad - i\sum_k \kappa_k r_k(0)e^{-i\omega_k t} \\
&= \frac{1}{i\hbar}[a, H_S] - \frac{1}{2}(L'/c)|\kappa(\omega_C)|^2 a - i\sum_k \kappa_k r_k(0)e^{-i\omega_k t} \\
&= \frac{1}{i\hbar}[a, H_S] - \kappa a - i\sum_k \kappa_k r_k(0)e^{-i\omega_k t}.
\end{aligned} \tag{1.71}$$

The last term on the right-hand side of (1.71) describes the driving of the cavity mode by the freely evolving modes of the reservoir field. The cavity mode will only respond to those free-field modes with frequencies close to ω_C. For these frequencies we may read (1.71) with $\kappa_k = \kappa(\omega_C) = -ie^{i(\phi_R-\phi_T)}\sqrt{c/L'}\sqrt{2\kappa}$, and (1.67b) with $\sqrt{\omega_k/\omega_C} = 1$. Thus, (1.71) may be written in the form

$$\dot{a} = \frac{1}{i\hbar}[a, H_S] - \kappa a - \sqrt{c/L'}\sqrt{2\kappa}\, r_f. \tag{1.72}$$

This equation allows us to express the correlations between the free field and the source field in terms of averages involving system operators alone. By multiplying (1.72) on the left or right by an arbitrary system operator $\hat{O}$, we find

$$\begin{aligned}&\sqrt{c/L'}\sqrt{2\kappa}\langle\hat{O}(t+\tau)r_f(t)\rangle\\ &\quad= \frac{1}{i\hbar}\langle\hat{O}(t+\tau)[a, H_S](t)\rangle - \kappa\langle\hat{O}(t+\tau)a(t)\rangle - \langle\hat{O}(t+\tau)\dot{a}(t)\rangle,\end{aligned} \tag{1.73a}$$

$$\begin{aligned}&\sqrt{c/L'}\sqrt{2\kappa}\langle r_f(t)\hat{O}(t+\tau)\rangle\\ &\quad= \frac{1}{i\hbar}\langle[a, H_S](t)\hat{O}(t+\tau)\rangle - \kappa\langle a(t)\hat{O}(t+\tau)\rangle - \langle\dot{a}(t)\hat{O}(t+\tau)\rangle,\end{aligned} \tag{1.73b}$$

and, for $\tau > 0$,

$$\begin{aligned}&\sqrt{c/L'}\sqrt{2\kappa}\langle\hat{O}(t)r_f(t+\tau)\rangle\\ &\quad= -\left(\frac{d}{d\tau} + \kappa\right)\langle\hat{O}(t)a(t+\tau)\rangle + \frac{1}{i\hbar}\langle\hat{O}(t)[a, H_S](t+\tau)\rangle,\end{aligned} \tag{1.74a}$$

$$\begin{aligned}&\sqrt{c/L'}\sqrt{2\kappa}\langle r_f(t+\tau)\hat{O}(t)\rangle\\ &\quad= -\left(\frac{d}{d\tau} + \kappa\right)\langle a(t+\tau)\hat{O}(t)\rangle + \frac{1}{i\hbar}\langle[a, H_S](t+\tau)\hat{O}(t)\rangle.\end{aligned} \tag{1.74b}$$

The rest of the calculation involves knowing how to evaluate the correlation functions that involve time derivatives on the right-hand sides of (1.73) and (1.74) [1.8]. For a cavity mode that obeys the master equation (1.47) this leads to the results

$$\sqrt{c/L'}\sqrt{2\kappa}\langle\hat{O}(t+\tau)r_f(t)\rangle = \begin{cases} 0 & \tau < 0 \\ \kappa\bar{n}\langle[\hat{O}(t+\tau), a(t)]\rangle & \tau = 0 \\ 2\kappa\bar{n}\langle[\hat{O}(t+\tau), a(t)]\rangle & \tau > 0, \end{cases} \tag{1.75a}$$

and

$$\sqrt{c/L'}\sqrt{2\kappa}\langle r_f(t)\hat{O}(t+\tau)\rangle = \begin{cases} 0 & \tau < 0 \\ \kappa(\bar{n}+1)\langle[\hat{O}(t+\tau), a(t)]\rangle & \tau = 0 \\ 2\kappa(\bar{n}+1)\langle[\hat{O}(t+\tau), a(t)]\rangle & \tau > 0. \end{cases} \tag{1.75b}$$

We can now evaluate all of the terms in (1.68) and (1.69) for a cavity mode radiating into a thermal reservoir. Using (1.75a) we have

$$g_{ss}^{(1)}(\tau) = \left(\langle\hat{\mathcal{E}}^\dagger\hat{\mathcal{E}}\rangle_{ss}\right)^{-1}\Big\{(c/L')\langle r_f^\dagger(0)r_f(\tau)\rangle + 2\kappa\Big[\lim_{t\to\infty}\langle a^\dagger(t)a(t+\tau)\rangle\Big]$$
$$+2\kappa\bar{n}\Big[\lim_{t\to\infty}\langle[a^\dagger(t),a(t+\tau)]\rangle\Big]\Big\}, \qquad (1.76)$$

with

$$\langle\hat{\mathcal{E}}^\dagger\hat{\mathcal{E}}\rangle_{ss} = (c/L')\langle r_f^\dagger r_f\rangle + 2\kappa\langle a^\dagger a\rangle_{ss} + 2\kappa\bar{n}\langle[a^\dagger,a]\rangle_{ss}$$
$$= (c/L')\langle r_f^\dagger r_f\rangle + 2\kappa\big(\langle a^\dagger a\rangle_{ss} - \bar{n}\big). \qquad (1.77)$$

In steady state the presence of the cavity should be invisible to a measurement made on the total reservoir field; effectively, the cavity mode is simply "absorbed" into the reservoir, becoming part of a slightly larger thermal equilibrium system. We have not yet seen how to calculate the system correlation functions that appear on the right-hand side of (1.76). But it should not be difficult to accept the results

$$\langle a^\dagger a\rangle_{ss} = \bar{n}, \qquad (1.78a)$$

and

$$\lim_{t\to\infty}\langle a^\dagger(t)a(t+\tau)\rangle = \bar{n}e^{-i\omega_C\tau}e^{-\kappa|\tau|}, \qquad (1.78b)$$
$$\lim_{t\to\infty}\langle[a^\dagger(t),a(t+\tau)]\rangle = -e^{-i\omega_C t}e^{-\kappa|\tau|}. \qquad (1.78c)$$

When these are substituted into (1.76) and (1.77) we see that the interference term $2\kappa\bar{n}\lim_{t\to\infty}\langle[a^\dagger(t),a(t+\tau)]\rangle$ between the free field and the source field cancels the source term $2\kappa\lim_{t\to\infty}\langle a^\dagger(t)a(t+\tau)\rangle$. Thus,

$$g_{ss}^{(1)}(\tau) = \left(\langle r_f^\dagger r_f\rangle\right)^{-1}\langle r_f^\dagger(0)r_f(\tau)\rangle. \qquad (1.79)$$

We have recovered the reservoir correlation function; thus, the correlation function for the total field – free field plus source field – is unaffected by the presence of the cavity which is what we expect for our thermal equilibrium example.

References

[1.1] I. R. Senitzky, *Phys. Rev.*, **119**, 670 (1960); **124**, 642 (1961).
[1.2] W. H. Louisell, *Quantum Statistical Properties of Radiation*, Wiley: New York, 1973, pp. 331ff.
[1.3] M. Sargent III, M. O. Scully, and W. E. Lamb, Jr., *Laser Physics*, Addison-Wesley: Reading, Massachusetts, 1974, pp. 265ff.
[1.4] F. Haake, *Z. Phys.* **224**, 353 (1969); *ibid.*, 365 (1969).

[1.5] F. Haake, "Statistical Treatment of Open Systems by Generalized Master Equations," in *Springer Tracts in Modern Physics*, Vol. 66, Springer: Berlin, 1973, pp. 117ff.
[1.6] W. C. Schieve and J. W. Middleton, *International J. Quant. Chem.*, Quantum Chemistry Symposium 11, 625 (1977).
[1.7] C. W. Gardiner and M. J. Collett, *Phys. Rev.* A **31**, 3761 (1985).
[1.8] H. J. Carmichael, *J. Opt. Soc. Am.* B **4**, 1588 (1987).

Lecture 2 – Master Equations and Sources II

A cavity mode driven by thermal light does not provide a very interesting example of an optical source. Indeed, it is not really a source at all since the total field observed at the output of the cavity is just the thermal field that drives the cavity. It is only for a non-thermal-equilibrium system that we will see a bright light, different from the surroundings, emitted by a source. The importance of the cavity mode calculation is that it provides one of the building blocks that we will use to construct more interesting sources. At the end of this lecture we will meet some examples of more interesting photoemissive sources. But first, most of the lecture will be devoted to a discussion of two-state atoms which provide another building block for the construction of more interesting sources. Excited atoms act as a source of radiation through spontaneous and stimulated emission. We are going to use the master equation approach from Sect. 1.2 to treat these processes for an atom in thermal equilibrium.

2.1 Two-state atoms

We consider two states of an atom, designated $|1\rangle$ and $|2\rangle$, having energies E_1 and E_2 with $E_1 < E_2$. Radiative transitions between $|1\rangle$ and $|2\rangle$ are allowed in the dipole approximation. Our objective is to describe energy dissipation and polarization damping through the coupling of the $|1\rangle \to |2\rangle$ transition to the many modes of the vacuum radiation field (a reservoir of harmonic oscillators). For simplicity we assume that there are no transitions between $|1\rangle$ and $|2\rangle$ and other states of the atom. The extension to multilevel atoms can be found in Louisell [2.1] and Haken [2.2]. A treatment for just two levels which corresponds closely to our own is given in Sargent, Scully and Lamb [2.3].

Any two-state system can be described in terms of the Pauli spin operators. We will be using this description in many of the following lectures and therefore we briefly review the relationship between these operators and quantities of physical interest such as the atomic inversion and polarization. A more complete coverage of this subject is given by Allen and Eberly [2.4].

If we have a representation in terms of a complete set of states $|n\rangle, n = 1, 2, \ldots$, any operator $\hat{O}$ can be expanded as

$$\hat{O} = \sum_{n,m} \langle n|\hat{O}|m\rangle |n\rangle\langle m|. \tag{2.1}$$

This follows after multiplying on the left and right by the identity operator $\hat{I} = \sum_n |n\rangle\langle n|$. The $\langle n|\hat{O}|m\rangle$ are the matrix representation of $\hat{O}$ with respect to the basis $|n\rangle$. If we adopt the energy eigenstates $|1\rangle$ and $|2\rangle$ as a basis for our two-state atom, the unperturbed atomic Hamiltonian H_A can be written in the form

$$\begin{aligned} H_A &= E_1|1\rangle\langle 1| + E_2|2\rangle\langle 2| \\ &= \tfrac{1}{2}(E_1 + E_2)\hat{I} + \tfrac{1}{2}(E_2 - E_1)\sigma_z, \end{aligned} \tag{2.2}$$

where

$$\sigma_z = |2\rangle\langle 2| - |1\rangle\langle 1|. \tag{2.3}$$

The first term in (2.2) is a constant and can be omitted if we refer atomic energies to the middle of the atomic transition. We then write

$$H_A = \tfrac{1}{2}\hbar\omega_A\sigma_z, \qquad \omega_A = (E_2 - E_1)/\hbar. \tag{2.4}$$

Consider now the dipole moment operator $e\hat{\boldsymbol{q}}$, where e is the electronic charge and $\hat{\boldsymbol{q}}$ is the coordinate operator for the bound electron:

$$\begin{aligned} e\hat{\boldsymbol{q}} &= e\sum_{n,m=1}^{2} \langle n|\hat{\boldsymbol{q}}|m\rangle|n\rangle\langle m| \\ &= e\big(\langle 1|\hat{\boldsymbol{q}}|2\rangle|1\rangle\langle 2| + \langle 2|\hat{\boldsymbol{q}}|1\rangle|2\rangle\langle 1|\big) \\ &= \boldsymbol{d}_{12}\sigma_- + \boldsymbol{d}_{21}\sigma_+, \end{aligned} \tag{2.5}$$

where we set $\langle 1|\hat{\boldsymbol{q}}|1\rangle = \langle 2|\hat{\boldsymbol{q}}|2\rangle = 0$, assuming atomic states whose symmetry guarantees zero permanent dipole moment, and we define the dipole matrix element

$$\boldsymbol{d}_{12} = e\langle 1|\hat{\boldsymbol{q}}|2\rangle, \qquad \boldsymbol{d}_{21} = (\boldsymbol{d}_{12})^*, \tag{2.6}$$

and operators

$$\sigma_- = |1\rangle\langle 2|, \quad \sigma_+ = |2\rangle\langle 1|. \tag{2.7}$$

The matrix representations for the operators introduced in (2.3) and (2.7) are

$$\sigma_z = \begin{pmatrix} 1 & 0 \\ 0 & -1 \end{pmatrix}, \qquad \sigma_- = \begin{pmatrix} 0 & 0 \\ 1 & 0 \end{pmatrix}, \qquad \sigma_+ = \begin{pmatrix} 0 & 1 \\ 0 & 0 \end{pmatrix}. \tag{2.8}$$

If we write

$$\sigma_\pm = \tfrac{1}{2}(\sigma_x \pm i\sigma_y), \tag{2.9}$$

with

$$\sigma_x = \begin{pmatrix} 0 & 1 \\ 1 & 0 \end{pmatrix}, \qquad \sigma_y = \begin{pmatrix} 0 & -i \\ i & 0 \end{pmatrix}, \tag{2.10}$$

then σ_x, σ_y, and σ_z are the *Pauli spin matrices* introduced initially in the context of magnetic transitions in spin-$\frac{1}{2}$ systems. In their application to two-state atoms σ_z, σ_-, and σ_+ are referred to as *pseudo*-spin operators, since, here, the two levels are not associated with the states of a real spin. From the relationships above it is straightforward to deduce the following:

1. the commutation relations

$$[\sigma_+, \sigma_-] = \sigma_z, \qquad [\sigma_\pm, \sigma_z] = \mp 2\sigma_\pm; \tag{2.11}$$

2. the action on atomic states:

$$\sigma_z|1\rangle = -|1\rangle, \qquad \sigma_z|2\rangle = |2\rangle, \tag{2.12a}$$

$$\sigma_-|1\rangle = 0, \qquad \sigma_-|2\rangle = |1\rangle, \tag{2.12b}$$

$$\sigma_+|1\rangle = |2\rangle, \qquad \sigma_+|2\rangle = 0. \tag{2.12c}$$

From (2.12b) and (2.12c) the common designation of σ_- and σ_+ as atomic lowering and raising operators is clear.

We will formulate our description of two-state atoms in terms of the operators σ_z, σ_-, and σ_+. For an atomic state specified by a density operator ρ, expectation values of σ_z, σ_-, and σ_+ are just the matrix elements of the density operator, and give the population difference

$$\langle \sigma_z \rangle = \mathrm{tr}(\rho\sigma_z) = \langle 2|\rho|2\rangle - \langle 1|\rho|1\rangle = \rho_{22} - \rho_{11}, \tag{2.13}$$

and the mean atomic polarization

$$\begin{aligned} \langle e\hat{\boldsymbol{q}} \rangle &= \boldsymbol{d}_{12}\mathrm{tr}(\rho\sigma_-) + \boldsymbol{d}_{21}\mathrm{tr}(\rho\sigma_+) \\ &= \boldsymbol{d}_{12}\langle 2|\rho|1\rangle + \boldsymbol{d}_{21}\langle 1|\rho|2\rangle \\ &= \boldsymbol{d}_{12}\,\rho_{21} + \boldsymbol{d}_{21}\,\rho_{12}. \end{aligned} \tag{2.14}$$

2.2 Master equation for a two-state atom in thermal equilibrium

We consider an atom which is radiatively damped by its interaction with the many modes of the radiation field in thermal equilibrium at temperature T. This field acts as a reservoir of harmonic oscillators. The reservoir is essentially the same as that considered in Lecture 1. However, the geometry is now different; electromagnetic field modes impinge on the atom from all directions in three-dimensional space, instead of entering an optical cavity by propagating in one dimension. Within the general formula for a system S

interacting with a reservoir R, the Hamiltonian (1.1) is given in the rotating-wave and dipole approximations by

$$H_S = \tfrac{1}{2}\hbar\omega_A\sigma_z, \tag{2.15a}$$

$$H_R = \sum_{\boldsymbol{k},\lambda} \hbar\omega_k r^\dagger_{\boldsymbol{k},\lambda} r_{\boldsymbol{k},\lambda}, \tag{2.15b}$$

$$H_{SR} = \sum_{\boldsymbol{k},\lambda} \hbar\left(\kappa^*_{\boldsymbol{k},\lambda} r^\dagger_{\boldsymbol{k},\lambda}\sigma_- + \kappa_{\boldsymbol{k},\lambda} r_{\boldsymbol{k},\lambda}\sigma_+\right), \tag{2.15c}$$

with

$$\kappa_{\boldsymbol{k},\lambda} = -ie^{i\boldsymbol{k}\cdot\boldsymbol{r}_A}\sqrt{\frac{\omega_k}{2\hbar\epsilon_0 V}}\,\hat{e}_{\boldsymbol{k},\lambda}\cdot\boldsymbol{d}_{21}. \tag{2.16}$$

The summation extends over reservoir oscillators (electromagnetic field modes) with wavevectors $\boldsymbol{k}$ and polarization states λ, and corresponding frequencies ω_k and unit polarization vectors $\hat{e}_{\boldsymbol{k},\lambda}$; the atom is positioned at $\boldsymbol{r}_A$ and V is the quantization volume. The general formalism from Sect. 1.2 now takes us directly to (1.18), where from (1.17) and (2.15) we must make the identification:

$$s_1 = \sigma_-, \qquad s_2 = \sigma_+, \tag{2.17a}$$

$$\Gamma_1 = \Gamma^\dagger = \sum_{\boldsymbol{k},\lambda} \kappa^*_{\boldsymbol{k},\lambda} r^\dagger_{\boldsymbol{k},\lambda}, \qquad \Gamma_2 = \Gamma = \sum_{\boldsymbol{k},\lambda} \kappa_{\boldsymbol{k},\lambda} r_{\boldsymbol{k},\lambda}. \tag{2.17b}$$

In the interaction picture,

$$\tilde{\Gamma}_1(t) = \tilde{\Gamma}^\dagger(t) = \sum_{\boldsymbol{k},\lambda} \kappa^*_{\boldsymbol{k},\lambda} r^\dagger_{\boldsymbol{k},\lambda} e^{i\omega_k t}, \tag{2.18a}$$

$$\tilde{\Gamma}_2(t) = \tilde{\Gamma}(t) = \sum_{\boldsymbol{k},\lambda} \kappa_{\boldsymbol{k},\lambda} r_{\boldsymbol{k},\lambda} e^{-i\omega_k t}, \tag{2.18b}$$

and

$$\tilde{s}_1(t) = e^{i(\omega_A\sigma_z/2)t}\sigma_- e^{-i(\omega_A\sigma_z/2)t} = \sigma_- e^{-i\omega_A t}, \tag{2.19a}$$

$$\tilde{s}_2(t) = e^{i(\omega_A\sigma_z/2)t}\sigma_+ e^{-i(\omega_A\sigma_z/2)t} = \sigma_+ e^{i\omega_A}. \tag{2.19b}$$

Aside from the obvious notational differences, (2.18) and (2.19) are the same as (1.25) and (1.24), respectively, with the substitution $a \to \sigma_-$, $a^\dagger \to \sigma_+$. The derivation of the master equation for a two-state atom then follows in complete analogy to the derivation of the master equation for the cavity mode, aside from two minor differences: (1) The explicit evaluation of the summation over reservoir oscillators now involves a summation over wavevector directions and polarization states. (2) Commutation relations used to reduce the master equation to its simplest form are now different. Neither of these steps are taken in passing from (1.18) to (1.33), or in evaluating the time integrals using (1.39). We can therefore simply make the substitution $a \to \sigma_-$, $a^\dagger \to \sigma_+$ in (1.36) to write

$$\begin{aligned}\dot{\tilde{\rho}} = &\left[\frac{\gamma}{2}(\bar{n}+1) + i(\Delta' + \Delta)\right](\sigma_-\tilde{\rho}\sigma_+ - \sigma_+\sigma_-\tilde{\rho}) + \text{h.c.} \\ &+ \left(\frac{\gamma}{2}\bar{n} + i\Delta'\right)(\sigma_+\tilde{\rho}\sigma_- - \tilde{\rho}\sigma_-\sigma_+) + \text{h.c.},\end{aligned} \tag{2.20}$$

with $\bar{n} = \bar{n}(\omega_A, T)$, and

$$\gamma = 2\pi \sum_\lambda \int d^3k\, g(\boldsymbol{k})|\kappa(\boldsymbol{k},\lambda)|^2 \delta(kc - \omega_A), \tag{2.21}$$

$$\Delta = \sum_\lambda P \int d^3k\, \frac{g(\boldsymbol{k})|\kappa(\boldsymbol{k},\lambda)|^2}{\omega_A - kc}, \tag{2.22}$$

$$\Delta' = \sum_\lambda P \int d^3k\, \frac{g(\boldsymbol{k})|\kappa(\boldsymbol{k},\lambda)|^2}{\omega_A - kc}\bar{n}(kc, T). \tag{2.23}$$

We have grouped the terms slightly differently in (2.20), but the correspondence to (1.36) is clear when we note that, there, $\alpha = \kappa + i\Delta$ and $\beta = \kappa\bar{n} + i\Delta'$. Equation (2.20) gives

$$\begin{aligned}\dot{\tilde{\rho}} &= \frac{\gamma}{2}(\bar{n}+1)(2\sigma_-\tilde{\rho}\sigma_+ - \sigma_+\sigma_-\tilde{\rho} - \tilde{\rho}\sigma_+\sigma_-) - i(\Delta' + \Delta)[\sigma_+\sigma_-, \tilde{\rho}] \\ &\quad + \frac{\gamma}{2}\bar{n}(2\sigma_+\tilde{\rho}\sigma_- - \sigma_-\sigma_+\tilde{\rho} - \tilde{\rho}\sigma_-\sigma_+) + i\Delta'[\sigma_-\sigma_+, \tilde{\rho}] \\ &= -i\frac{1}{2}(2\Delta' + \Delta)[\sigma_z, \tilde{\rho}] + \frac{\gamma}{2}(\bar{n}+1)(2\sigma_-\tilde{\rho}\sigma_+ - \sigma_+\sigma_-\tilde{\rho} - \tilde{\rho}\sigma_+\sigma_-) \\ &\quad + \frac{\gamma}{2}\bar{n}(2\sigma_+\tilde{\rho}\sigma_- - \sigma_-\sigma_+\tilde{\rho} - \tilde{\rho}\sigma_-\sigma_+),\end{aligned} \tag{2.24}$$

where we have used

$$\sigma_+\sigma_- = |2\rangle\langle 1|1\rangle\langle 2| = |2\rangle\langle 2| = \tfrac{1}{2}(1 + \sigma_z), \tag{2.25a}$$

$$\sigma_-\sigma_+ = |1\rangle\langle 2|2\rangle\langle 1| = |1\rangle\langle 1| = \tfrac{1}{2}(1 - \sigma_z). \tag{2.25b}$$

Finally, transforming back to the Schrödinger picture using (1.46), we obtain the *master equation for a two-state atom in thermal equilibrium:*

$$\begin{aligned}\dot{\rho} = &-i\tfrac{1}{2}\omega_A'[\sigma_z, \rho] + \frac{\gamma}{2}(\bar{n}+1)(2\sigma_-\rho\sigma_+ - \sigma_+\sigma_-\rho - \rho\sigma_+\sigma_-) \\ &+ \frac{\gamma}{2}\bar{n}(2\sigma_+\rho\sigma_- - \sigma_-\sigma_+\rho - \rho\sigma_-\sigma_+),\end{aligned} \tag{2.26}$$

with

$$\omega_A' = \omega_A + 2\Delta' + \Delta. \tag{2.27}$$

The symmetric grouping of terms we have adopted identifies a transition rate from $|2\rangle \to |1\rangle$, described by the term proportional to $(\gamma/2)(\bar{n}+1)$, and a transition rate from $|1\rangle \to |2\rangle$, described by the term proportional to $(\gamma/2)\bar{n}$. The former contains a rate for spontaneous transitions, independent of $\bar{n}$, and a rate for stimulated transitions induced by thermal photons, proportional to $\bar{n}$; the latter gives a rate for absorptive transitions which take thermal photons from the equilibrium electromagnetic field. Notice the frequency shift $\omega_A' - \omega_A$. This is the Lamb shift, including a temperature-dependent contribution $2\Delta'$, which did not appear for the harmonic oscillator. The appearance of the temperature-dependent piece here is a consequence of the commutator $[\sigma_-, \sigma_+] = -\sigma_z$ in place of the corresponding $[a, a^\dagger] = 1$ for the harmonic oscillator. From (2.22), (2.23), and (1.31)

$$\begin{aligned}2\Delta' + \Delta &= \sum_\lambda P \int d^3k \, \frac{g(\boldsymbol{k})|\kappa(\boldsymbol{k},\lambda)|^2}{\omega_A - kc}[1 + 2\bar{n}(kc, T)] \\ &= \sum_\lambda P \int d^3k \, \frac{g(\boldsymbol{k})|\kappa(\boldsymbol{k},T)|^2}{\omega_A - kc} \coth\left(\frac{\hbar kc}{2k_B T}\right),\end{aligned} \tag{2.28}$$

where k_B is Boltzmann's constant. The temperature independent term in the square bracket gives the normal *Lamb shift*, while the term proportional to $2\bar{n}$ gives the frequency shift induced via the ac Stark effect by the thermal reservoir field [2.5, 2.6, 2.7]. We might note that the use of the rotating-wave approximation in our calculation does not give the correct nonrelativistic result for the Lamb shift [2.8]. In place of $(\omega_A - kc)^{-1}$ in (2.22) and (2.23) there should be $(\omega_A - kc)^{-1} + (\omega_A + kc)^{-1}$. After making this replacement it can be shown that (2.23) gives the formula for the *temperature-dependent shift* derived in Ref. [2.6]:

$$2\Delta' = \frac{1}{4\pi\epsilon_0}\frac{4d_{12}^2}{3\hbar\pi c^3} \, P\int_0^\infty d\omega\, \omega^3 \left(\frac{1}{\omega_A - \omega} + \frac{1}{\omega_A + \omega}\right) \frac{1}{e^{\hbar\omega/k_B T} - 1}. \tag{2.29}$$

The corresponding formula for the *Lamb shift* is

$$\Delta = \frac{1}{4\pi\epsilon_0}\frac{2d_{12}^2}{3\hbar\pi c^3} \, P\int_0^\infty d\omega\, \omega^3 \left(\frac{1}{\omega_A - \omega} + \frac{1}{\omega_A + \omega}\right). \tag{2.30}$$

If we have a correct description of spontaneous emission we must expect the damping constant γ appearing in (2.26) to give the correct result for the Einstein A coefficient. We can check this by performing the integration over wavevectors and the polarization summation in (2.21).

Adopting spherical coordinates in $\boldsymbol{k}$-space, the density of states for each polarization state λ is given by

$$g(\boldsymbol{k})d^3k = \frac{\omega^2 V}{8\pi^3 c^3} d\omega \sin\theta d\theta d\phi. \tag{2.31}$$

Substituting (2.31) and (2.16) into (2.21),

$$\begin{aligned}\gamma &= 2\pi \sum_\lambda \int_0^\infty d\omega \int_0^\pi \sin\theta d\theta \int_0^{2\pi} d\phi \, \frac{\omega^2 V}{8\pi^3 c^3} \frac{\omega}{2\hbar\epsilon_0 V} (\hat{e}_{\boldsymbol{k},\lambda} \cdot \boldsymbol{d}_{12})^2 \delta(\omega - \omega_A) \\ &= \frac{\omega_A^3}{8\pi^2 \epsilon_0 \hbar c^3} \sum_\lambda \int_0^\pi \sin\theta d\theta \int_0^{2\pi} d\phi \, (\hat{e}_{\boldsymbol{k},\lambda} \cdot \boldsymbol{d}_{12})^2.\end{aligned} \tag{2.32}$$

Now, for each $\boldsymbol{k}$ we can choose polarization states λ_1 and λ_2 so that the first polarization state gives $\hat{e}_{\boldsymbol{k},\lambda_1} \cdot \boldsymbol{d}_{12} = 0$. This is achieved with the geometry illustrated in Fig. 2.1. Then for the second polarization state

$$(\hat{e}_{\boldsymbol{k},\lambda_2} \cdot \boldsymbol{d}_{12})^2 = d_{12}^2(1 - \cos^2 \alpha) = d_{12}^2\left[1 - (\hat{d}_{12} \cdot \hat{k})^2\right], \tag{2.33}$$

where $\hat{d}_{12}$ and $\hat{k}$ are unit vectors in the directions of $\boldsymbol{d}_{12}$ and $\boldsymbol{k}$, respectively. The angular integrals are now easily performed if we choose the k_z-axis to correspond to the $\hat{d}_{12}$ direction. We have

$$\begin{aligned}\int_0^\pi \sin\theta d\theta \int_0^{2\pi} d\phi\, (\hat{e}_{\boldsymbol{k},\lambda_2} \cdot \boldsymbol{d}_{12})^2 &= d_{12}^2 \int_0^{2\pi} d\phi \int_0^\pi d\theta \sin\theta(1 - \cos^2\theta) \\ &= \frac{8\pi}{3} d_{12}^2 .\end{aligned} \tag{2.34}$$

From (2.32) and (2.34)

$$\gamma = \frac{1}{4\pi\epsilon_0} \frac{4\omega_A^3 d_{12}^2}{3\hbar c^3}. \tag{2.35}$$

This is the correct result for the *Einstein A coefficient*, as obtained from the Wigner-Weisskopf theory of natural linewidth [2.9, 2.10].

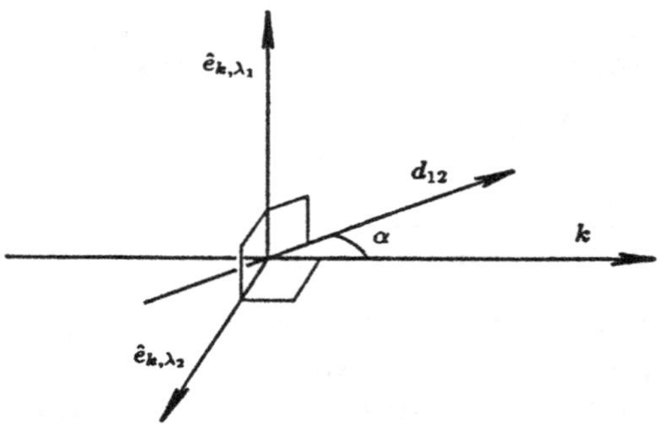

Fig. 2.1. Geometry of polarization states chosen for evaluating (2.32).

2.3 Phase destroying processes

The interaction (2.15c) with the many mode electromagnetic field causes both energy loss from the atom and damping of the atomic polarization. Polarization damping results from a randomization of the phases of the atomic wavefunctions by thermal and vacuum fluctuations in the electromagnetic field, which causes the overlap of the upper and lower state wavefunctions to decay in time. It is often necessary to account for additional dephasing interactions; these might arise from elastic collisions in an atomic vapor, or elastic phonon scattering in a solid. Let us see what terms are added to the master equation to describe such processes.

A phenomenological model describing atomic dephasing can be obtained by adding two further reservoir interactions to the Hamiltonian (2.15). We add

$$H_{dephase} = H_{R_1} + H_{R_2} + H_{SR_1} + H_{SR_2}, \tag{2.36}$$

with

$$H_{R_1} + H_{R_2} = \sum_j \hbar\omega_{1j}\, r_{1j}^\dagger r_{1j} + \sum_j \hbar\omega_{2j}\, r_{2j}^\dagger r_{2j}, \tag{2.37a}$$

$$H_{SR_1} + H_{SR_2} = \sum_{j,k} \hbar\kappa_{1jk}\, r_{1j}^\dagger r_{1k}\, \sigma_-\sigma_+ + \sum_{j,k} \hbar\kappa_{2jk}\, r_{2j}^\dagger r_{2k}\, \sigma_+\sigma_-. \tag{2.37b}$$

Of course, these additional reservoirs are not associated with additional radiated fields; they simply modify the dynamics of the radiating source. The complete reservoir seen by the atom is now composed of three subsystems: $R = R_{12} \oplus R_1 \oplus R_2$, where R_{12} is the reservoir defined by (2.15b). These reservoir subsystems are assumed to be statistically independent, with the density operator R_0 given by the product of three thermal equilibrium operators in the form of (1.22). The interactions H_{SR_1} and H_{SR_2} describe the scattering of quanta from the atom while it is in states $|1\rangle$ and $|2\rangle$, respectively; they sum over virtual processes which scatter quanta with energies $\hbar\omega_{1k}$ and $\hbar\omega_{2k}$ into quanta with energies $\hbar\omega_{1j}$ and $\hbar\omega_{2j}$ while leaving the state of the atom unchanged.

The terms which are added to the master equation by these new reservoir interactions follow in a rather straightforward manner from the general form (1.18) for the master equation in the Born approximation. In addition to the reservoir operators $\tilde{\Gamma}_1(t)$ and $\tilde{\Gamma}_2(t)$ which are defined by the interaction with R_{12} [Eqs. (2.18)], we must introduce operators $\tilde{\Gamma}_3(t)$ and $\tilde{\Gamma}_4(t)$ to account for the interactions with R_1 and R_2. However, we first have to take care of a problem, one which was not met in deriving master equations for the cavity mode and the radiatively damped atom. Equation (1.18) was obtained using the assumption that all reservoir operators coupling to the system S have zero mean in the state R_0 [below (1.12)]. This is not true for the reservoir operators coupling to $\sigma_-\sigma_+$ and $\sigma_+\sigma_-$ in (2.37b); terms with $j = k$ in the summation over reservoir modes have nonzero averages proportional to mean thermal occupation numbers. To overcome this difficulty the interaction between S and the mean reservoir "field" can be included in H_S rather than H_{SR}. With the use of (2.25), in place of (2.37a) and (2.37b) we may write

$$H_S = \tfrac{1}{2}\hbar(\omega_A + \delta_p)\sigma_z, \tag{2.38}$$

and

$$H_{SR_1}+H_{SR_2}$$
$$=\sum_{j,k}\hbar\kappa_{1jk}(r_{1j}^{\dagger}r_{1k}-\delta_{jk}\bar{n}_{1j})\sigma_-\sigma_+ +\sum_{j,k}\hbar\kappa_{2jk}(r_{2j}^{\dagger}r_{2k}-\delta_{jk}\bar{n}_{2j})\sigma_+\sigma_-,\tag{2.39}$$

with the frequency shift δ_p given by

$$\begin{aligned}\delta_p&=\sum_j(\kappa_{2jj}\bar{n}_{2j}-\kappa_{1jj}\bar{n}_{1j})\\&=\int_0^{\infty}d\omega\big[g_2(\omega)\kappa_2(\omega,\omega)-g_1(\omega)\kappa_1(\omega,\omega)\big]\bar{n}(\omega,T).\end{aligned}\tag{2.40}$$

$\bar{n}_{1j}=\bar{n}(\omega_{1j},T)$ and $\bar{n}_{2j}=\bar{n}(\omega_{2j},T)$ are mean occupation numbers for reservoir modes with frequencies ω_{1j} and ω_{2j}, respectively, and in (2.40) the summation over reservoir modes has been converted to an integration by introducing the densities of states $g_1(\omega)$ and $g_2(\omega)$. The sum of (2.38) and (2.39) gives the same Hamiltonian as the sum of (2.37a) and (2.37b); but now the reservoir operators appearing in H_{SR_1} and H_{SR_2} have zero mean.

We may now proceed directly from (1.18). After transforming to the interaction picture, the interaction Hamiltonian (2.39) is written in the form of (1.17) with

$$\tilde{s}_3(t)=\sigma_-\sigma_+,\tag{2.41a}$$
$$\tilde{s}_4(t)=\sigma_+\sigma_-,\tag{2.41b}$$

and

$$\tilde{\Gamma}_3(t)=\sum_{j,k}\kappa_{1jk}\left(r_{1j}^{\dagger}r_{1k}\,e^{i(\omega_{1j}-\omega_{1k})t}-\delta_{jk}\bar{n}_{1j}\right),\tag{2.42a}$$
$$\tilde{\Gamma}_4(t)=\sum_{j,k}\kappa_{2jk}\left(r_{2j}^{\dagger}r_{2k}\,e^{i(\omega_{2j}-\omega_{2k})t}-\delta_{jk}\bar{n}_{2j}\right).\tag{2.42b}$$

These are to be substituted – together with $\tilde{s}_1(t)$, $\tilde{s}_2(t)$, $\tilde{\Gamma}_1(t)$, and $\tilde{\Gamma}_2(t)$ from (2.18) and (2.19) – into (1.18). Since the reservoir subsystems are statistically independent and all reservoir operators have zero mean, all of the cross terms involving correlation functions for products of operators from different reservoir subsystems will vanish. Thus, the spontaneous and stimulated emission terms arising from the interaction with $\tilde{\Gamma}_1$ and $\tilde{\Gamma}_2$ are obtained exactly as in Sect. 2.2. The additional terms from the interaction with $\tilde{\Gamma}_3$ and $\tilde{\Gamma}_4$ take the form

$$\begin{aligned}\left(\dot{\tilde{\rho}}\right)_{dephase} = -\int_0^t dt' & [\sigma_-\sigma_+\sigma_-\sigma_+\tilde{\rho}(t') - \sigma_-\sigma_+\tilde{\rho}(t')\sigma_-\sigma_+]\langle\tilde{\Gamma}_3(t)\tilde{\Gamma}_3(t')\rangle_{R_1} \\ & + [\tilde{\rho}(t')\sigma_-\sigma_+\sigma_-\sigma_+ - \sigma_-\sigma_+\tilde{\rho}(t')\sigma_-\sigma_+]\langle\tilde{\Gamma}_3(t')\tilde{\Gamma}_3(t)\rangle_{R_1} \\ & + [\sigma_+\sigma_-\sigma_+\sigma_-\tilde{\rho}(t') - \sigma_+\sigma_-\tilde{\rho}(t')\sigma_+\sigma_-]\langle\tilde{\Gamma}_4(t)\tilde{\Gamma}_4(t')\rangle_{R_2} \\ & + [\tilde{\rho}(t')\sigma_+\sigma_-\sigma_+\sigma_- - \sigma_+\sigma_-\tilde{\rho}(t')\sigma_+\sigma_-]\langle\tilde{\Gamma}_4(t')\tilde{\Gamma}_4(t)\rangle_{R_2}.\end{aligned} \tag{2.43}$$

We will evaluate the first of the reservoir correlation functions appearing in (2.43) from which the others follow in a similar form. From (2.42a),

$$\begin{aligned}&\langle\tilde{\Gamma}_3(t)\tilde{\Gamma}_3(t')\rangle_{R_1} \\ &= \mathrm{tr}\Bigg[R_{10}\sum_{j,k}\sum_{j',k'}\kappa_{1jk}\,\kappa_{1j'k'}\left(r_{1j}^\dagger r_{1k}e^{i(\omega_{1j}-\omega_{1k})t} - \delta_{jk}\bar{n}_{1j}\right) \\ &\qquad\times\left(r_{1j'}^\dagger r_{1k'}e^{i(\omega_{1j'}-\omega_{1k'})t'} - \delta_{j'k'}\bar{n}_{1j'}\right)\Bigg] \\ &= \mathrm{tr}\Bigg[R_{10}\Bigg(\sum_{j,k}\sum_{j',k'}\kappa_{1jk}\,\kappa_{1j'k'}\,r_{1j}^\dagger r_{1k}\,r_{1j'}^\dagger r_{1k'}\,e^{i(\omega_{1j}-\omega_{1k})t}e^{i(\omega_{1j'}-\omega_{1k'})t'} \\ &\quad-\sum_{j}\sum_{j',k'}\kappa_{1jj}\,\kappa_{1j'k'}\,\bar{n}_{1j}\,r_{1j'}^\dagger r_{1k'}\,e^{i(\omega_{1j'}-\omega_{1k'})t'} \\ &\quad-\sum_{j,k}\sum_{j'}\kappa_{1jk}\,\kappa_{1j'j'}\,r_{1j}^\dagger r_{1k}\,\bar{n}_{1j'}\,e^{i(\omega_{1j}-\omega_{1k})t}\Bigg)\Bigg] \\ &\quad+\sum_{j}\sum_{j'}\kappa_{1jj}\,\kappa_{1j'j'}\,\bar{n}_{1j}\,\bar{n}_{1j'},\end{aligned}$$

where R_{10} is the thermal equilibrium density operator for the reservoir subsystem R_1. The nonvanishing contributions to the trace are now as follows: the first double sum contributes for $j = k \neq j' = k'$, for $j = k' \neq k = j'$, and for $j = k = j' = k'$; the second double sum contributes for $j' = k'$; and the third double sum for $j = k$. The correlation function becomes

$$\begin{aligned}&\langle\tilde{\Gamma}_3(t)\tilde{\Gamma}_3(t')\rangle_{R_1} \\ &= \sum_{\substack{j,j'\\ j\neq j'}}\kappa_{1jj}\,\kappa_{1j'j'}\,\bar{n}_{1j}\,\bar{n}_{1j'} + \sum_{\substack{j,j'\\ j\neq j'}}\kappa_{1jj'}\,\kappa_{1j'j}\,\bar{n}_{1j}(\bar{n}_{1j'}+1)e^{i(\omega_{1j}-\omega_{1j'})(t-t')} \\ &\quad+\sum_j\kappa_{1jj}^2\overline{n_{1j}^2} - 2\sum_{j,j'}\kappa_{1jj}\,\kappa_{1j'j'}\,\bar{n}_{1j}\,\bar{n}_{1j'} + \sum_{j,j'}\kappa_{1jj}\,\kappa_{1j'j'}\,\bar{n}_{1j}\,\bar{n}_{1j'},\end{aligned}$$

where the first three terms come from the first double sum, and the fourth term comes from the second and third double sums. Noting that $\overline{n_{1j}^2} = \bar{n}_{1j}^2 + \bar{n}_{1j}(\bar{n}_{1j}+1)$, we see that the sums for $j \neq j'$ are completed for all j and j' by the third term in this expression; setting $\kappa_{1jj'}\kappa_{1j'j} = |\kappa_{1jj'}|^2$ – required for (2.37b) to be Hermitian – we arrive at the result

$$\langle \tilde{\Gamma}_3(t)\tilde{\Gamma}_3(t')\rangle_{R_1} = \sum_{j,j'} |\kappa_{1jj'}|^2 \bar{n}_{1j}(\bar{n}_{1j'}+1)e^{i(\omega_{1j}-\omega_{1j'})(t-t')}. \tag{2.44a}$$

Similar expressions follow for the other reservoir correlation functions:

$$\langle \tilde{\Gamma}_4(t)\tilde{\Gamma}_4(t')\rangle_{R_2} = \sum_{j,j'} |\kappa_{2jj'}|^2 \bar{n}_{2j}(\bar{n}_{2j'}+1)e^{i(\omega_{2j}-\omega_{2j'})(t-t')}, \tag{2.44b}$$

and

$$\langle \tilde{\Gamma}_3(t')\tilde{\Gamma}_3(t)\rangle_{R_1} = \left(\langle \tilde{\Gamma}_3(t)\tilde{\Gamma}_3(t')\rangle_{R_1}\right)^*, \tag{2.44c}$$

$$\langle \tilde{\Gamma}_4(t')\tilde{\Gamma}_4(t)\rangle_{R_2} = \left(\langle \tilde{\Gamma}_4(t)\tilde{\Gamma}_4(t')\rangle_{R_2}\right)^*. \tag{2.44d}$$

If reservoir correlation times are very short compared to the time scale for the system dynamics, the time integral in (2.43) can be treated in the same fashion as in Sect. 1.3. After simplifying the operator products using (2.25), (2.43) then gives

$$\left(\dot{\tilde{\rho}}\right)_{dephase} = -i\tfrac{1}{2}\Delta_p[\sigma_z,\tilde{\rho}] + \frac{\gamma_p}{2}(\sigma_z\tilde{\rho}\sigma_z - \tilde{\rho}), \tag{2.45}$$

with

$$\begin{aligned}\gamma_p = \pi\int_0^\infty d\omega\,&\left[g_2(\omega)^2|\kappa_2(\omega,\omega)|^2 + g_1(\omega)^2|\kappa_1(\omega,\omega)|^2\right]\\ &\times \bar{n}(\omega,T)\left[\bar{n}(\omega,T)+1\right],\end{aligned} \tag{2.46}$$

$$\begin{aligned}\Delta_p = P\int_0^\infty d\omega\int_0^\infty d\omega'\,&\frac{g_2(\omega)g_2(\omega')|\kappa_2(\omega,\omega')|^2 - g_1(\omega)g_1(\omega')|\kappa_1(\omega,\omega')|^2}{\omega-\omega'}\\ &\times \bar{n}(\omega,T).\end{aligned} \tag{2.47}$$

We add (2.45) to the terms describing radiative damping given by (2.24), and transform back to the Schrödinger picture using (1.46) and (2.38) to obtain the *master equation for a two-state atom in thermal equilibrium with nonradiative dephasing*:

$$\begin{aligned}\dot{\rho} = &-i\tfrac{1}{2}\omega_A'[\sigma_z,\rho] + \frac{\gamma}{2}(\bar{n}+1)(2\sigma_-\rho\sigma_+ - \sigma_+\sigma_-\rho - \rho\sigma_+\sigma_-)\\ &+\frac{\gamma}{2}\bar{n}(2\sigma_+\rho\sigma_- - \sigma_-\sigma_+\rho - \rho\sigma_-\sigma_+) + \frac{\gamma_p}{2}(\sigma_z\rho\sigma_z - \rho),\end{aligned} \tag{2.48}$$

where the shifted atomic frequency is now

$$\omega_A' = \omega_A + 2\Delta' + \Delta + \delta_p + \Delta_p, \tag{2.49}$$

with $2\Delta'+\Delta$, δ_p, and Δ_p given by (2.28), (2.40), and (2.47).

2.4 The radiated field

As we saw in Lecture 1, the master equation approach focuses first on the dynamics of the source – in this case the two-state atom. We are ultimately interested, however, in the properties of the field radiated by the source. We therefore need a relationship analogous to that derived in Sect. 1.4 between source operators and the radiated field:

$$\hat{\boldsymbol{E}}(\boldsymbol{r},t) = \hat{\boldsymbol{E}}^{(+)}(\boldsymbol{r},t) + \hat{\boldsymbol{E}}^{(-)}(\boldsymbol{r},t), \tag{2.50a}$$

where

$$\hat{\boldsymbol{E}}^{(+)}(\boldsymbol{r},t) = i\sum_{\boldsymbol{k},\lambda}\sqrt{\frac{\hbar\omega_k}{2\epsilon_0 V}}\,\hat{e}_{\boldsymbol{k},\lambda} r_{\boldsymbol{k},\lambda}(t) e^{i\boldsymbol{k}\cdot\boldsymbol{r}}, \tag{2.50b}$$

$$\hat{\boldsymbol{E}}^{(-)}(\boldsymbol{r},t) = \hat{\boldsymbol{E}}^{(+)}(\boldsymbol{r},t)^{\dagger}. \tag{2.50c}$$

As before we will separate this field into a freely evolving part and a contribution from the source using the Heisenberg equations of motion for the reservoir modes.

The Heisenberg equations of motion give

$$\dot{r}_{\boldsymbol{k},\lambda} = -i\omega_k r_{\boldsymbol{k},\lambda} - i\kappa^*_{\boldsymbol{k},\lambda}\sigma_-. \tag{2.51}$$

If we write

$$r_{\boldsymbol{k},\lambda} = \tilde{r}_{\boldsymbol{k},\lambda} e^{-i\omega_k t}, \tag{2.52a}$$

$$\sigma_- = \tilde{\sigma}_- e^{-i\omega_A t}, \tag{2.52b}$$

after formal integration of (2.51)

$$\tilde{r}_{\boldsymbol{k},\lambda}(t) = r_{\boldsymbol{k},\lambda}(0) - i\kappa^*_{\boldsymbol{k},\lambda}\int_0^t dt'\,\tilde{\sigma}_-(t') e^{i(\omega_k-\omega_A)t'}. \tag{2.53}$$

Separation of the rapidly oscillating term in (2.52b) is motivated by the solution of Heisenberg equations for a free atom [Eqs. (2.19)]. Now, substituting for $r_{\boldsymbol{k},\lambda}(t)$ in (2.50a) and introducing the explicit form of the coupling constant from (2.16), the field operator becomes

$$\hat{\boldsymbol{E}}^{(+)}(\boldsymbol{r},t) = \hat{\boldsymbol{E}}_f^{(+)}(\boldsymbol{r},t) + \hat{\boldsymbol{E}}_s^{(+)}(\boldsymbol{r},t), \tag{2.54}$$

with

$$\hat{\boldsymbol{E}}_f^{(+)}(\boldsymbol{r},t) = i\sum_{\boldsymbol{k},\lambda}\sqrt{\frac{\hbar\omega_k}{2\epsilon_0 V}}\,\hat{e}_{\boldsymbol{k},\lambda} r_{\boldsymbol{k},\lambda}(0) e^{-i(\omega_k t-\boldsymbol{k}\cdot\boldsymbol{r})}, \tag{2.55}$$

and

$$\hat{\boldsymbol{E}}_s^{(+)}(\boldsymbol{r},t) = i\frac{1}{2\epsilon_0 V}e^{-i\omega_A t}\sum_{\boldsymbol{k},\lambda}\omega_k \hat{e}_{\boldsymbol{k},\lambda}(\hat{e}_{\boldsymbol{k},\lambda}\cdot \boldsymbol{d}_{12})e^{i\boldsymbol{k}\cdot(\boldsymbol{r}-\boldsymbol{r}_A)}$$
$$\times \int_0^t dt'\,\tilde{\sigma}_-(t')e^{i(\omega_k-\omega_A)(t'-t)}. \tag{2.56}$$

Here $\hat{\boldsymbol{E}}_f^{(+)}(\boldsymbol{r},t)$ describes the free evolution of the electromagnetic field in the absence of the atom; $\hat{\boldsymbol{E}}_s^{(+)}(\boldsymbol{r},t)$ is a source field radiated by the atom. It remains to perform the summation and integration in (2.56).

The summation over $\boldsymbol{k}$ is performed by introducing the density of states (2.31):

$$\hat{\boldsymbol{E}}_s^{(+)}(\boldsymbol{r},t) = i\frac{1}{16\pi^3\epsilon_0 c^3}e^{-i\omega_A t}\sum_\lambda \int_0^\infty d\omega \int_0^\pi \sin\theta\, d\theta \int_0^{2\pi} d\phi$$
$$\times\, \omega^3 \hat{e}_{\boldsymbol{k},\lambda}(\hat{e}_{\boldsymbol{k},\lambda}\cdot \boldsymbol{d}_{12})e^{i(\omega r/c)\cos\theta}\int_0^t dt'\,\tilde{\sigma}_-(t')e^{i(\omega-\omega_A)(t'-t)}, \tag{2.57}$$

where we have chosen a geometry with the origin in $\boldsymbol{r}$-space at the site of the atom and the k_z-axis in the direction of $\boldsymbol{r}$. One polarization state may be chosen perpendicular to both $\boldsymbol{k}$ and $\boldsymbol{d}_{12}$, as in Fig. 2.1, and for the second we can write

$$\hat{e}_{\boldsymbol{k},\lambda_2}(\hat{e}_{\boldsymbol{k},\lambda_2}\cdot \boldsymbol{d}_{12}) = -\hat{e}_{\boldsymbol{k},\lambda_2} d_{12}\sin\alpha = -(\boldsymbol{d}_{12}\times\hat{k})\times\hat{k}, \tag{2.58}$$

where $\hat{k}$ is a unit vector in the direction of $\boldsymbol{k}$. Setting

$$\hat{k} = \hat{r}\cos\theta + \hat{k}_x\sin\theta\cos\phi + \hat{k}_y\sin\theta\sin\phi, \tag{2.59}$$

where $\hat{k}_x$, $\hat{k}_y$, and $\hat{r} = \boldsymbol{r}/r$ are unit vectors along the Cartesian axes in $\boldsymbol{k}$-space, the angular integrals are readily evaluated to give

$$\hat{\boldsymbol{E}}_s^{(+)}(\boldsymbol{r},t) = \frac{1}{8\pi^2\epsilon_0 c^2 r}(\boldsymbol{d}_{12}\times\hat{r})\times\hat{r}\int_0^\infty d\omega\,\omega^2\Bigg[e^{-i\omega_A(t+r/c)}$$
$$\times\int_0^t dt'\,\tilde{\sigma}_-(t')e^{i(\omega-\omega_A)(t'-t-r/c)} - e^{-i\omega_A(t-r/c)}$$
$$\times\int_0^t dt'\,\tilde{\sigma}_-(t')e^{i(\omega-\omega_A)(t'-t+r/c)}\Bigg]. \tag{2.60}$$

Now, since we have removed the rapid oscillation at the atomic resonance frequency by the transformation (2.52b), $\tilde{\sigma}_-$ is expected to vary slowly in comparison with the optical period – on a time scale characterized by $\gamma^{-1}\sim 10^{-8}$s (for optical frequencies), while $\omega_A^{-1}\sim 10^{-15}$s. Thus, for frequencies outside the range $-100\gamma \lesssim \omega-\omega_A \lesssim 100\gamma$, say, the time integrals in (2.60) average to zero. This means that over the important range of the frequency integral $\omega^2 \approx \omega_A^2 + 2(\omega-\omega_A)\omega_A$ varies by less than 0.01% from $\omega^2=\omega_A^2$.

We therefore replace ω^2 by ω_A^2 and extend the frequency integral to $-\infty$. We then find

$$\begin{aligned}
&\hat{\boldsymbol{E}}_s^{(+)}(\boldsymbol{r},t)\\
&= \frac{\omega_A^2}{4\pi\epsilon_0 c^2 r}(\boldsymbol{d}_{12}\times\hat{r})\times\hat{r}\left[e^{-i\omega_A(t+r/c)}\int_0^t dt'\,\tilde{\sigma}_-(t')\delta(t'-t-r/c)\right.\\
&\quad\left.-e^{-i\omega_A(t-r/c)}\int_0^t dt'\,\tilde{\sigma}_-(t')\delta(t'-t+r/c)\right]\\
&= -\frac{\omega_A^2}{4\pi\epsilon_0 c^2 r}(\boldsymbol{d}_{12}\times\hat{r})\times\hat{r}\,\sigma_-(t-r/c).
\end{aligned} \tag{2.61}$$

This is precisely the familiar result for classical dipole radiation with the dipole moment *operator* $\boldsymbol{d}_{12}\sigma_-$ in place of the classical dipole moment.

2.5 Other sources: resonance fluorescence, lasers, parametric oscillators

The issue raised in Sect. 1.5 regarding correlations between the free field and source field is relevant again for the atomic source. However, we have seen what this issue is in principle and we will not spend time on the specific details of the correlations for the atomic case. In fact, in the case of an atomic source, the occasion for which we really need these correlations will be even rarer than it is for cavity-based sources. The reason is that the scattering from an atom goes into a 4π solid angle. Even if the atom is illuminated by a non-vacuum field, it is unlikely that the illuminating field will fill all 4π of the modes seen by the atom; and the scattered light will generally be viewed from a direction that is outside the solid angle filled by the illuminating field – a direction in which the free field is in the vacuum state. Of course, one example where this is not so is the example of thermal equilibrium, which is intrinsically isotropic. But the thermal equilibrium calculations are just an introduction; they are not what really interests us. At optical frequencies and laboratory temperatures the thermal photon number $\bar{n}$ is completely negligible ($\sim 10^{-42}$). We will therefore omit the terms proportional to $\bar{n}$ in most of the examples discussed in later lectures. We should remember, however, that thermal effects are not negligible at microwave frequencies where even at liquid helium temperatures a few thermal photons are present. This regime is quite relevant to current research because of the work on micromasers and cavity Q.E.D. [2.11, 2.12].

Once we have understood the derivation of the two master equations (1.47) and (2.26) [and perhaps (2.48)] we can quickly write down master equations for a variety of sources that involve cavity modes, atoms, and their interaction. To conclude this lecture let us see a few of the examples we will be using in later lectures.

With minor modification (2.26) is converted into the *master equation for resonance fluorescence:*

$$\dot{\rho} = -i\tfrac{1}{2}\omega_A[\sigma_z,\rho] + i(\Omega/2)\left[e^{-i\omega_A t}\sigma_+ + e^{i\omega_A t}\sigma_-,\rho\right] \\ + \frac{\gamma}{2}(2\sigma_-\rho\sigma_+ - \sigma_+\sigma_-\rho - \rho\sigma_+\sigma_-). \tag{2.62}$$

All we have done here is add the second commutator term on the right-hand side to describe the interaction, in the dipole and rotating-wave approximations, of the two-state atom with a resonant laser field. Because the driving laser is modeled by a highly populated field mode that is essentially undepleted by its interaction with the atom, its amplitude may be treated as a c-number rather than as an operator. To be more precise, the parameter $\Omega = 2dE/\hbar$ is the *Rabi frequency* associated with the driving field amplitude E; d is the projection of the atomic dipole moment on the polarization direction of the driving field.

Now to an example involving cavity modes. The master equation (1.47) provides the basic building block for the quantum mechanical treatment of various nonlinear optical models. One important example is the degenerate parametric oscillator. This system involves two cavity modes, a pump mode of frequency $2\omega_C$ and a subharmonic mode of frequency ω_C, coupled by a $\chi^{(2)}$ nonlinearity. The pump mode is driven by a classical field injected into the cavity and the output of the cavity is a source of the subharmonic. The *master equation for the degenerate parametric oscillator* has the form

$$\dot{\rho} = -i\omega_C[a^\dagger a,\rho] - i2\omega_C[b^\dagger b,\rho] \\ + (g/2)[a^{\dagger 2}b - a^2 b^\dagger,\rho] - i[\bar{\mathcal{E}}_i e^{-i2\omega_C t}b^\dagger + \bar{\mathcal{E}}_i^* e^{i2\omega_C t}b,\rho] \\ + \kappa(2a\rho a^\dagger - a^\dagger a\rho - \rho a^\dagger a) + \kappa_p(2b\rho b^\dagger - b^\dagger b\rho - \rho b^\dagger b). \tag{2.63}$$

Here $a^\dagger$ and a are creation and annihilation operators for the subharmonic, and $b^\dagger$ and b are creation and annihilation operators for the pump; g is a coupling coefficient proportional to $\chi^{(2)}$, $\bar{\mathcal{E}}_i$ is proportional to the amplitude of the field driving the pump mode; and κ and κ_p are decay rates (half-widths) for the subharmonic and pump modes, respectively. The master equation (2.63) is comprised of four commutators coming from the $(1/i\hbar)[H_S,\rho]$ in (1.46) and two decay terms associated with the loss of energy into cavity output fields at frequencies ω_C and $2\omega_C$. Often a simpler version of (2.63) can be used, with the pump field entering only as a parameter. We will meet this simpler equation and the reasons justifying the simplification later on.

Finally, just one rather complicated example – the *master equation for the single-mode homogeneously-broadened laser with atomic dephasing*:

$$\dot{\rho} = -i\tfrac{1}{2}\omega_C[J_z,\rho] - i\omega_C[a^\dagger a,\rho] + g[a^\dagger J_- - aJ_+,\rho] \\ + \kappa(2a\rho a^\dagger - a^\dagger a\rho - \rho a^\dagger a)$$

$$+\frac{\gamma_\downarrow}{2}\left(\sum_{j=1}^{N} 2\sigma_{j-}\rho\sigma_{j+} - \sigma_{j+}\sigma_{j-}\rho - \rho\sigma_{j+}\sigma_{j-}\right)$$
$$+\frac{\gamma_\uparrow}{2}\left(\sum_{j=1}^{N} 2\sigma_{j+}\rho\sigma_{j-} - \sigma_{j-}\sigma_{j+}\rho - \rho\sigma_{j-}\sigma_{j+}\right)$$
$$+\frac{\gamma_p}{2}\left(\sum_{j=1}^{N} \sigma_{jz}\rho\sigma_{jz} - N\rho\right). \tag{2.64}$$

We are not going to discuss the analysis of this more complicated system in future lectures; but is worthwhile just stating the master equation to see how simple its structure really is. The first three terms on the right-hand side obviously describe N identical two-state atoms interacting on resonance with a cavity mode; the operators J_+, J_-, and J_z are sums over the σ_{j+}, σ_{j-}, and σ_{jz} for N atoms. The next term describes the decay from the cavity mode (laser output field); it is given by the master equation (1.47). The last three terms describe the radiative decay, incoherent pumping, and dephasing of the N lasing atoms: the first of these comes from the master equation (2.26), the third is the dephasing term added in (2.48), and the second – the pumping term – is the decay term in (2.26) written backwards – σ_{j-} and σ_{j+} operators are interchanged to make the "spontaneous emission" go from the lower state to the upper state instead of the reverse.

References

[2.1] W. H. Louisell, *Quantum Statistical Properties of Radiation*, Wiley: New York, 1973, pp. 347ff.

[2.2] H. Haken, *Handbuch der Physik*, Vol. XXV/2c, ed. by L. Genzel, Springer: Berlin, 1970, pp. 27ff.

[2.3] M. Sargent III, M. O. Scully, and W. E. Lamb, Jr., *Laser Physics*, Addison-Wesley: Reading, Massachusetts, 1974, pp. 273ff.

[2.4] L. Allen and J. H. Eberly, *Optical Resonance and Two-Level Atoms*, Wiley: New York, 1975, pp. 28ff.

[2.5] T. F. Gallagher and W. E. Cook, *Phys. Rev. Lett.* **42**, 835 (1979).

[2.6] J. W. Farley and W. H. Wing, *Phys. Rev.* A **23**, 5 (1981).

[2.7] L. Hollberg and J. L. Hall, *Phys. Rev. Lett.* **53**, 230 (1984).

[2.8] G. S. Agarwal, *Phys. Rev.* A **4**, 1778 (1971); *Phys. Rev.* A **7**, 1195 (1973).

[2.9] Reference [2.1], pp. 250ff.

[2.10] V. G. Weisskopf and E. Wigner, *Z. Phys.* **63**, 54 (1930).

[2.11] S. Haroche and J. M. Raimond, "Radiative Properties of Rydberg States in Resonant Cavities," in *Advances in Atomic and Molecular Physics*,

Vol. 20, Eds. D. Bates and B. Bederson, Academic Press: New York, 1985, pp. 347ff.
[2.12] D. Meschede, H. Walther, and G. Müller, *Phys. Rev. Lett.* **54**, 551 (1985).

Lecture 3 – Standard Methods of Analysis I

We have now seen how to obtain a master equation to describe the quantum dynamics of a photoemissive source. What we need next are methods for analyzing this equation. In the next two lectures we will review some of the standard methods for doing this. On the way we will not only pick up analytical tools, we also treat a number of examples that introduce us to some important physical results concerning the spectra and photon statistics of photoemissive sources.

3.1 Operator expectation values

We begin by returning to the master equation (1.47) for a cavity mode driven by thermal light. The cavity mode should be damped through the loss of energy into its radiated field. Let us make some simple checks to see if the master equation describes the damped evolution we expect. Since we have formulated our theory in the Schrödinger picture we cannot obtain solutions for operators themselves, only for their expectation values. For example, if we multiply (1.47) on the left by a and take the trace (over the system S) we obtain an equation for $\langle a\rangle = \mathrm{tr}(a\rho)$:

$$\begin{aligned}\langle\dot{a}\rangle &= -i\omega_C\,\mathrm{tr}(aa^\dagger a\rho - a\rho a^\dagger a) + \kappa\,\mathrm{tr}(2a^2\rho a^\dagger - aa^\dagger a\rho - a\rho a^\dagger a)\\ &\quad + 2\kappa\bar{n}\,\mathrm{tr}(a^2\rho a^\dagger + aa^\dagger\rho a - aa^\dagger a\rho - a\rho a a^\dagger)\\ &= -i\omega_C\,\mathrm{tr}\big[(aa^\dagger - a^\dagger a)a\rho\big] + \kappa\,\mathrm{tr}\Big[(a^\dagger a - aa^\dagger)a\rho\Big]\\ &\quad + 2\kappa\bar{n}\,\mathrm{tr}\big[(a^\dagger a - aa^\dagger)a\rho + a(aa^\dagger - a^\dagger a)\rho\big]\\ &= -(\kappa + i\omega_C)\langle a\rangle,\end{aligned} \tag{3.1}$$

where we have used the cyclic property of the trace and the boson commutation relation. Equation (3.1) does describe the decay of the mean mode amplitude we expected. In a similar way we obtain

$$\langle\dot{\hat{n}}\rangle = -2\kappa(\langle\hat{n}\rangle - \bar{n}), \tag{3.2}$$

with the solution

$$\langle\hat{n}(t)\rangle = \langle\hat{n}(0)\rangle e^{-2\kappa t} + \bar{n}(1 - e^{-2\kappa t}). \tag{3.3}$$

Notice how thermal fluctuations are fed into the cavity from the reservoir; the mean energy does not decay to zero but to the mean energy for an harmonic oscillator with frequency ω_C in thermal equilibrium at temperature T.

Equations (3.1) and (3.2) are examples of operator expectation value equations – the simplest way to get physical information from a master equation. We can obtain equivalent equations for the atomic source described by the master equation (2.26). Since $\langle\sigma_z\rangle$, $\langle\sigma_-\rangle$, and $\langle\sigma_+\rangle$ are simply related to the matrix elements of ρ, one way to proceed in this case is to take the matrix elements of (2.26) directly. This gives

$$\dot\rho_{22} = -\gamma(\bar n+1)\rho_{22} + \gamma\bar n\rho_{11}, \tag{3.4a}$$

$$\dot\rho_{11} = -\gamma\bar n\rho_{11} + \gamma(\bar n+1)\rho_{22}, \tag{3.4b}$$

$$\dot\rho_{21} = -\left[\frac{\gamma}{2}(2\bar n+1) + i\omega_A\right]\rho_{21}, \tag{3.4c}$$

$$\dot\rho_{12} = -\left[\frac{\gamma}{2}(2\bar n+1) - i\omega_A\right]\rho_{12}. \tag{3.4d}$$

Equations (3.4a) and (3.4b) are the *Einstein rate equations*; they clearly illustrate the physical interpretation of the two terms – proportional to $(\gamma/2)(\bar n+1)$ and $(\gamma/2)\bar n$ – in the master equation; the former describes $|2\rangle \to |1\rangle$ transitions at a rate $\gamma(\bar n+1)$ and the latter describes $|1\rangle \to |2\rangle$ transitions at a rate $\gamma\bar n$. In the steady state the balance between upwards and downwards transitions leads to a thermal distribution between states $|1\rangle$ and $|2\rangle$.

Using the relations $\langle\sigma_z\rangle = \rho_{22} - \rho_{11}$, $\langle\sigma_-\rangle = \rho_{21}$, $\langle\sigma_+\rangle = \rho_{12}$, and $\rho_{11} + \rho_{22} = 1$, (3.4a)–(3.4d) can be written as operator expectation value equations. If we include the coherent driving term that appears in the master equation for resonance fluorescence [Eq. (2.62)] we obtain

$$\langle\dot\sigma_-\rangle = -\left[\frac{\gamma}{2}(2\bar n+1) + i\omega_A\right]\langle\sigma_-\rangle - i(\Omega/2)e^{-i\omega_A t}\langle\sigma_z\rangle, \tag{3.5a}$$

$$\langle\dot\sigma_+\rangle = -\left[\frac{\gamma}{2}(2\bar n+1) - i\omega_A\right]\langle\sigma_+\rangle + i(\Omega/2)e^{i\omega_A t}\langle\sigma_z\rangle, \tag{3.5b}$$

$$\langle\dot\sigma_z\rangle = -\gamma\big[\langle\sigma_z\rangle(2\bar n+1) + 1\big] + i\Omega e^{-i\omega_A t}\langle\sigma_+\rangle - i\Omega e^{i\omega_A t}\langle\sigma_-\rangle. \tag{3.5c}$$

These are the *optical Bloch equations* with radiative damping, so called for their relationship to the equations of a spin-$\frac{1}{2}$ system in a magnetic field. As we noted at the end of the last lecture, at optical frequencies and laboratory temperatures $\bar n$ can be set to zero. If we also neglect the effects of spontaneous decay, which is valid for short times, the optical Bloch equations are equivalent to the classical equations for a magnetic moment $\boldsymbol{m}$ in a rotating magnetic field $\boldsymbol{B}$; with $\langle\sigma_x\rangle$ and $\langle\sigma_y\rangle$ defined as in (2.9), we can write

$$\dot{\boldsymbol{m}} = \boldsymbol{B}\times\boldsymbol{m}, \tag{3.6}$$

with

$$\boldsymbol{m} = \langle\sigma_x\rangle\hat{x} + \langle\sigma_y\rangle\hat{y} + \langle\sigma_z\rangle\hat{z}, \tag{3.7}$$

and

$$\boldsymbol{B} = -(\Omega\cos\omega_A t)\hat{x} - (\Omega\sin\omega_A t)\hat{y} + \omega_A\hat{z}, \tag{3.8}$$

where $\hat{x}, \hat{y}$, and $\hat{z}$ are orthogonal unit vectors.

An idea of the dynamics contained in the optical Bloch equations is obtained from their solution for the initial state $|1\rangle$ ($\bar{n} = 0$):

$$\begin{aligned} e^{\pm i\omega_A t}\langle\sigma_\mp(t)\rangle = &\pm i\frac{1}{\sqrt{2}}\frac{Y}{1+Y^2}\left[1 - e^{-(3\gamma/4)t}\left(\cosh\delta t + \frac{(3\gamma/4)}{\delta}\sinh\delta t\right)\right] \\ &\pm i\sqrt{2}Ye^{-(3\gamma/4)t}\frac{(\gamma/4)}{\delta}\sinh\delta t, \end{aligned} \tag{3.9}$$

$$\langle\sigma_z(t)\rangle = -\frac{1}{1+Y^2}\left[1 + Y^2e^{-(3\gamma/4)t}\left(\cosh\delta t + \frac{(3\gamma/4)}{\delta}\sinh\delta t\right)\right], \tag{3.10}$$

where

$$Y = \frac{\sqrt{2}\Omega}{\gamma}, \tag{3.11}$$

$$\delta = \sqrt{\left(\frac{\gamma}{4}\right)^2 - \Omega^2} = \frac{\gamma}{4}\sqrt{1-8Y^2}. \tag{3.12}$$

We will make use of this solution shortly to derive some of the properties of the fluorescence from a two-state atom. But first we need to make a diversion to consider one other piece of formalism.

3.2 Correlation functions: the quantum regression theorem

We have developed a formalism which allows us, in principle, to solve for the density operator for a system (source) interacting with a reservoir. From this density operator we can obtain time-dependent expectation values for any operator acting in the Hilbert space of the system S. What, however, about products of operators evaluated at two different times? Of particular interest, for example, are the first-order and second-order correlation functions of the field radiated by the source. For a cavity mode with the reservoir in the vacuum state (see Sects. 1.4 and 1.5) these are given by

$$\begin{aligned} G^{(1)}(t, t+\tau) &\propto \langle a^\dagger(t)a(t+\tau)\rangle, \\ G^{(2)}(t, t+\tau) &\propto \langle a^\dagger(t)a^\dagger(t+\tau)a(t+\tau)a(t)\rangle. \end{aligned}$$

The first-order correlation function is required for calculating the spectrum of the field. The second-order correlation function gives information about the photon statistics and describes photon bunching or antibunching.

Clearly, averages involving two times cannot be calculated directly from the master equation – at least, not without a little extra thought. We need to return to the microscopic picture of system plus reservoir. At this level two-time averages are defined in the usual way in the Heisenberg representation. Our objective, then, is to derive a relationship which allows us to calculate these averages at the macroscopic level using the master equation for the reduced density operator alone; thus, in some approximate way we wish to carry out the trace over reservoir variables explicitly, as we did in deriving the master equation. The result we obtain is known as the quantum regression theorem and is attributed to Lax [3.1, 3.2].

Recall our microscopic formulation of system S coupled to reservoir R. The Hamiltonian for the composite system $S \oplus R$ takes the form given in (1.1). The density operator is designated $\chi(t)$ and satisfies Schrödinger's equation (1.4). Our derivation of the master equation gives us an equation for the reduced density operator (1.15), which we will now write formally as

$$\dot{\rho} = \mathcal{L}\rho, \tag{3.13}$$

where $\mathcal{L}$ is a generalized Liouvillian – $\mathcal{L}$ operates on operators rather than states; for example, for the cavity mode driven by thermal light the action of $\mathcal{L}$ on an arbitrary operator $\hat{O}$ is defined by the equation

$$\begin{aligned} \mathcal{L}\hat{O} = & -i\omega_C[a^\dagger a, \hat{O}] + \kappa(2a\hat{O}a^\dagger - a^\dagger a\hat{O} - \hat{O}a^\dagger a) \\ & + 2\kappa\bar{n}(a\hat{O}a^\dagger + a^\dagger\hat{O}a - a^\dagger a\hat{O} - \hat{O}aa^\dagger). \end{aligned} \tag{3.14}$$

Now, within the microscopic formalism multi-time averages are straightforwardly defined in the Heisenberg picture. In particular, the average of a product of operators evaluated at two different times is given by

$$\langle \hat{O}_1(t)\hat{O}_2(t')\rangle = \mathrm{tr}_{S\oplus R}[\chi(0)\hat{O}_1(t)\hat{O}_2(t')], \tag{3.15}$$

where $\hat{O}_1$ and $\hat{O}_2$ are any two system operators. These operators satisfy the Heisenberg equations of motion

$$\dot{\hat{O}}_1 = \frac{1}{i\hbar}[\hat{O}_1, H], \tag{3.16a}$$

$$\dot{\hat{O}}_2 = \frac{1}{i\hbar}[\hat{O}_2, H], \tag{3.16b}$$

with the formal solutions

$$\hat{O}_1(t) = e^{(i/\hbar)Ht}\hat{O}_1(0)e^{-(i/\hbar)Ht}, \tag{3.17a}$$

$$\hat{O}_2(t') = e^{(i/\hbar)Ht'}\hat{O}_2(0)e^{-(i/\hbar)Ht'}. \tag{3.17b}$$

From (1.4) the formal solution for χ gives

$$\chi(0) = e^{(i/\hbar)Ht}\chi(t)e^{-(i/\hbar)Ht}. \tag{3.18}$$

We then substitute these solutions into (3.15) and use the cyclic property of the trace to obtain

$$\begin{aligned}\langle \hat{O}_1(t)\hat{O}_2(t')\rangle &= \mathrm{tr}_{S\oplus R}\left[e^{(i/\hbar)Ht}\chi(t)\hat{O}_1(0)e^{(i/\hbar)H(t'-t)}\hat{O}_2(0)e^{-(i/\hbar)Ht'}\right] \\ &= \mathrm{tr}_S\Big\{\hat{O}_2(0)\mathrm{tr}_R\Big[e^{-(i/\hbar)H(t'-t)}\chi(t)\hat{O}_1(0)e^{(i/\hbar)H(t'-t)}\Big]\Big\}.\end{aligned} \tag{3.19}$$

In the final step we use the fact that $\hat{O}_2$ is an operator in the Hilbert space of S alone.

We will now specialize to the case $t' \geq t$ and define

$$\tau = t' - t, \tag{3.20}$$

$$\chi_{\hat{O}_1}(\tau) = e^{-(i/\hbar)H\tau}\chi(t)\hat{O}_1(0)e^{(i/\hbar)H\tau}. \tag{3.21}$$

Clearly, $\chi_{\hat{O}_1}$ satisfies the equation

$$\frac{d\chi_{\hat{O}_1}}{d\tau} = \frac{1}{i\hbar}\left[H, \chi_{\hat{O}_1}\right] \tag{3.22}$$

with

$$\chi_{\hat{O}_1}(0) = \chi(t)\hat{O}_1(0). \tag{3.23}$$

If we are to eliminate explicit reference to the reservoir in (3.19) we need to evaluate the reservoir trace over $\chi_{\hat{O}_1}(\tau)$ to obtain the reduced operator

$$\rho_{\hat{O}_1}(\tau) = \mathrm{tr}_R\left[\chi_{\hat{O}_1}(\tau)\right], \tag{3.24}$$

where

$$\rho_{\hat{O}_1}(0) = \mathrm{tr}_R[\chi(t)\hat{O}_1(0)] = \mathrm{tr}_R[\chi(t)]\hat{O}_1(0) = \rho(t)\hat{O}_1(0); \tag{3.25}$$

notice that $\rho_{\hat{O}_1}(\tau)$ is the term $\mathrm{tr}_R[\cdots]$ inside the curly brackets in (3.19). If we assume $\chi(t)$ factorizes as $\rho(t)R_0$ in the spirit of (1.13), we can write, from (3.23) and (3.25),

$$\chi_{\hat{O}_1}(0) = R_0[\rho(t)\hat{O}_1(0)] = R_0\,\rho_{\hat{O}_1}(0). \tag{3.26}$$

Equations (3.22), (3.24), and (3.26) are now equivalent to (1.4), (1.2), and (1.10) – namely, to the starting equations in our derivation of the master equation. We can find an equation for $\rho_{\hat{O}_1}(\tau)$ in the Born-Markoff approximation following a completely analogous course to that followed in deriving the master equation. Since (1.4) and (3.22) contain the same Hamiltonian H, in the formal notation of (3.13) we will arrive at the equation

$$\frac{d\rho_{\hat{O}_1}}{d\tau} = \mathcal{L}\rho_{\hat{O}_1}, \tag{3.27}$$

whose solution is

$$\rho_{\hat{O}_1}(\tau) = e^{\mathcal{L}\tau}\left[\rho_{\hat{O}_1}(0)\right] = e^{\mathcal{L}\tau}[\rho(t)\hat{O}_1(0)]. \tag{3.28}$$

When we substitute for $\rho_{\hat{O}_1}(\tau)$ in (3.19), we have ($\tau \geq 0$)

$$\langle \hat{O}_1(t)\hat{O}_2(t+\tau)\rangle = \mathrm{tr}_S\{\hat{O}_2(0)e^{\mathcal{L}\tau}[\rho(t)\hat{O}_1(0)]\}. \tag{3.29}$$

Following the same procedure we find ($\tau \geq 0$)

$$\langle \hat{O}_1(t+\tau)\hat{O}_2(t)\rangle = \mathrm{tr}_S\{\hat{O}_1(0)e^{\mathcal{L}\tau}[\hat{O}_2(0)\rho(t)]\}. \tag{3.30}$$

To calculate a correlation function $\langle \hat{O}_1(t)\hat{O}_2(t')\hat{O}_3(t)\rangle$ we cannot use (3.29) and (3.30) because noncommuting operators do not allow the reordering necessary to bring $\hat{O}_1(t)$ next to $\hat{O}_3(t)$. We may, however, generalize the approach taken above quite readily. Specifically, we have

$$\begin{aligned}
&\langle \hat{O}_1(t)\hat{O}_2(t')\hat{O}_3(t)\rangle \\
&\quad = \mathrm{tr}_{S\oplus R}\Big[e^{(i/\hbar)Ht}\chi(t)\hat{O}_1(0)e^{(i/\hbar)H(t'-t)}\hat{O}_2(0)e^{-(i/\hbar)H(t'-t)} \\
&\qquad \times\hat{O}_3(0)e^{-(i/\hbar)Ht}\Big] \\
&\quad = \mathrm{tr}_S\Big\{\hat{O}_2(0)\mathrm{tr}_R\Big[e^{-(i/\hbar)H(t'-t)}\hat{O}_3(0)\chi(t)\hat{O}_1(0)e^{(i/\hbar)H(t'-t)}\Big]\Big\}.
\end{aligned} \tag{3.31}$$

Defining

$$\chi_{\hat{O}_3\hat{O}_1}(\tau) = e^{-(i/\hbar)H\tau}\hat{O}_3(0)\chi(t)\hat{O}_1(0)e^{(i/\hbar)H\tau} \tag{3.32}$$

and

$$\rho_{\hat{O}_3\hat{O}_1}(\tau) = \mathrm{tr}_R\left[\chi_{\hat{O}_3\hat{O}_1}(\tau)\right] \tag{3.33}$$

as analogues of (3.21) and (3.24), we can proceed as we did above to the result ($\tau \geq 0$)

$$\langle \hat{O}_1(t)\hat{O}_2(t+\tau)\hat{O}_3(t)\rangle = \mathrm{tr}_S\{\hat{O}_2(0)e^{\mathcal{L}\tau}[\hat{O}_3(0)\rho(t)\hat{O}_1(0)]\}. \tag{3.34}$$

Equations (3.29) and (3.30) are, in fact, just special cases of (3.34) with either $\hat{O}_1(t)$ or $\hat{O}_3(t)$ set equal to the unit operator.

It is possible to work directly with the rather formal expressions derived above. However, these expressions can also be reduced to a more familiar form – a form which is perhaps more convenient for doing calculations [3.1]. Essentially, we will show that the equations of motion for expectation values of system operators (one-time averages), such as the optical Bloch equations, are also the equations of motion for correlation functions (two-time averages).

We begin by assuming that there exists a complete set of system operators $\hat{A}_\mu$, $\mu = 1, 2, \ldots$, in the following sense: that for an arbitrary operator $\hat{O}$, and for each $\hat{A}_\mu$,

$$\mathrm{tr}_S[\hat{A}_\mu(\mathcal{L}\hat{O})] = \sum_\lambda M_{\mu\lambda}\mathrm{tr}_S(\hat{A}_\lambda\hat{O}), \tag{3.35}$$

where the $M_{\mu\lambda}$ are constants. In particular, from this it follows that

$$\langle\dot{\hat{A}}_\mu\rangle = \mathrm{tr}_S(\hat{A}_\mu\dot{\rho}) = \mathrm{tr}_S[\hat{A}_\mu(\mathcal{L}\rho)] = \sum_\lambda M_{\mu\lambda}\langle\hat{A}_\lambda\rangle. \tag{3.36}$$

Thus, expectation values $\langle\hat{A}_\mu\rangle$, $\mu = 1, 2, \ldots$, obey a coupled set of linear equations with the evolution matrix M defined by the $M_{\mu\lambda}$ that appear in (3.35). In vector notation,

$$\langle\dot{\hat{\boldsymbol{A}}}\rangle = \boldsymbol{M}\langle\hat{\boldsymbol{A}}\rangle, \tag{3.37}$$

where $\hat{\boldsymbol{A}}$ is the column vector of operators $\hat{A}_\mu$, $\mu = 1, 2, \ldots$. Now, using (3.29) and (3.35) ($\tau \geq 0$):

$$\begin{aligned}\frac{d}{d\tau}\langle\hat{O}_1(t)\hat{A}_\mu(t+\tau)\rangle &= \mathrm{tr}_S\{\hat{A}_\mu(0)\big(\mathcal{L}e^{\mathcal{L}\tau}[\rho(t)\hat{O}_1(0)]\big)\}\\ &= \sum_\lambda M_{\mu\lambda}\mathrm{tr}_S\{\hat{A}_\lambda(0)e^{\mathcal{L}\tau}[\rho(t)\hat{O}_1(0)]\}\\ &= \sum_\lambda M_{\mu\lambda}\langle\hat{O}_1(t)\hat{A}_\lambda(t+\tau)\rangle,\end{aligned} \tag{3.38}$$

or,

$$\frac{d}{d\tau}\langle\hat{O}_1(t)\hat{\boldsymbol{A}}(t+\tau)\rangle = \boldsymbol{M}\langle\hat{O}_1(t)\hat{\boldsymbol{A}}(t+\tau)\rangle, \tag{3.39}$$

where $\hat{O}_1$ can be any system operator, not necessarily one of the $\hat{A}_\mu$. This result is just what would be obtained by removing the angular brackets from (3.37) (written with $t \to t+\tau$, and $\cdot \equiv d/dt \to d/d\tau$), multiplying on the left by $\hat{O}_1(t)$, and then replacing the angular brackets. Hence, for each operator $\hat{O}_1$, the set of correlation functions $\langle\hat{O}_1(t)\hat{A}_\mu(t+\tau)\rangle$, $\mu = 1, 2, \ldots$, with $\tau \geq 0$, satisfies the same equations (as functions of τ) as do the averages $\langle\hat{A}_\mu(t+\tau)\rangle$.

For $\tau \geq 0$ we can show, in a similar way, that

$$\frac{d}{d\tau}\langle\hat{\boldsymbol{A}}(t+\tau)\hat{O}_2(t)\rangle = \boldsymbol{M}\langle\hat{\boldsymbol{A}}(t+\tau)\hat{O}_2(t)\rangle. \tag{3.40}$$

Thus, we can also multiply (3.37) on the right by $\hat{O}_2(t)$, inside the average. We also find

$$\frac{d}{d\tau}\langle\hat{O}_1(t)\hat{\boldsymbol{A}}(t+\tau)\hat{O}_2(t)\rangle = \boldsymbol{M}\langle\hat{O}_1(t)\hat{\boldsymbol{A}}(t+\tau)\hat{O}_2(t)\rangle. \tag{3.41}$$

Perhaps this form of the quantum regression theorem seems restricted since its derivation relies on the existence of a set of operators $\hat{A}_\mu$, $\mu = 1, 2, \ldots$, for which (3.35) holds. But this is always so if a discrete basis $|n\rangle$, $n = 1, 2, \ldots$, exists; although in general the complete set of operators may be very large. Consider the operators

$$\hat{A}_\mu = \hat{A}_{nm} = |n\rangle\langle m|. \tag{3.42}$$

Then it is not difficult to show that

$$\mathrm{tr}_S[\hat{A}_{nm}(\mathcal{L}\hat{O})] = \sum_{n',m'} M_{nm;n'm'}\,\mathrm{tr}_S(\hat{A}_{n'm'}\hat{O}), \tag{3.43}$$

with

$$M_{nm;n'm'} = \langle m|\Big(\mathcal{L}|m'\rangle\langle n'|\Big)|n\rangle. \tag{3.44}$$

This is an expansion in the form of (3.35). The complete set of operators includes all the outer products $|n\rangle\langle m|$, $n = 1, 2, \ldots$, $m = 1, 2, \ldots$; this may be a small number of operators, a large, but finite, number of operators, or a double infinity of operators.

3.3 Optical spectra

Armed with the quantum regression theorem we are now able to get a lot more information out of the master equation for a photoemissive source. The first thing we might calculate is the spectrum of the source. To see how the calculations proceed we first consider a simple example based on the operator expectation value equations for the cavity mode driven by thermal light. We must calculate the correlation function $\langle a^\dagger(t)a(t+\tau)\rangle$. Equation (3.1) gives the equation of motion for the mean oscillator amplitude, and with $\hat{A}_1 = a$ and $\hat{O}_1 = a^\dagger$, from (3.37) and (3.39), we may write

$$\frac{d}{d\tau}\langle a^\dagger(t)a(t+\tau)\rangle = -(\kappa + i\omega_C)\langle a^\dagger(t)a(t+\tau)\rangle. \tag{3.45}$$

Thus,

$$\begin{aligned}\langle a^\dagger(t)a(t+\tau)\rangle &= \langle a^\dagger(t)a(t)\rangle e^{-(\kappa+i\omega_C)\tau}\\ &= \big[\langle\hat{n}(0)\rangle e^{-2\kappa t} + \bar{n}(1-e^{-2\kappa t})\big]e^{-(\kappa+i\omega_C)\tau},\end{aligned} \tag{3.46}$$

where the last line follows from (3.3). In the long-time (stationary) limit, the Fourier transform of the correlation function,

$$\langle a^\dagger(0)a(\tau)\rangle_{ss} \equiv \lim_{t\to\infty}\langle a^\dagger(t)a(t+\tau)\rangle = \bar{n}e^{-(\kappa+i\omega_C)\tau}, \tag{3.47}$$

gives the spectrum of the radiation from the cavity. This is clearly a Lorentzian with full-width (at half-maximum) 2κ.

Actually, we have to be cautious about calling this the spectrum of the radiation from the cavity, because we are neglecting the free field contributions discussed in Sects. 1.4 and 1.5. We can call the Fourier transform of (3.47) the spectrum of the radiation from the cavity if we refer to the radiation through a second mirror that is not illuminated by thermal light (Fig. 1.1) – it is the spectrum of filtered thermal light.

For a second example we calculate the *spectrum of spontaneous emission* from a two-state atom. In this case we start with the operator expectation value equations (3.5) with $\bar{n} = \Omega = 0$; we write these in vector form as

$$\langle \dot{\boldsymbol{s}} \rangle = \boldsymbol{M} \langle \boldsymbol{s} \rangle, \tag{3.48}$$

with

$$\boldsymbol{s} = \begin{pmatrix} \sigma_- \\ \sigma_+ \\ \sigma_+\sigma_- \end{pmatrix}, \tag{3.49}$$

$$\boldsymbol{M} = \operatorname{diag}\left[-\left(\frac{\gamma}{2} + i\omega_A\right), -\left(\frac{\gamma}{2} - i\omega_A\right), -\gamma\right]. \tag{3.50}$$

For $\tau \geq 0$, equations for nine correlation functions are obtained from (3.39):

$$\frac{d}{d\tau}\langle \sigma_-(t)\boldsymbol{s}(t+\tau)\rangle = \boldsymbol{M}\langle \sigma_-(t)\boldsymbol{s}(t+\tau)\rangle, \tag{3.51a}$$

$$\frac{d}{d\tau}\langle \sigma_+(t)\boldsymbol{s}(t+\tau)\rangle = \boldsymbol{M}\langle \sigma_+(t)\boldsymbol{s}(t+\tau)\rangle, \tag{3.51b}$$

$$\frac{d}{d\tau}\langle \sigma_+(t)\sigma_-(t)\boldsymbol{s}(t+\tau)\rangle = \boldsymbol{M}\langle \sigma_+(t)\sigma_-(t)\boldsymbol{s}(t+\tau)\rangle. \tag{3.51c}$$

With the atom prepared in its excited state, the initial condition is $\langle \sigma_- \rangle = \langle \sigma_+ \rangle = 0$, $\langle \sigma_+\sigma_- \rangle = \rho_{22} = 1$, and the solution to (3.48) is

$$\langle \boldsymbol{s} \rangle = \begin{pmatrix} 0 \\ 0 \\ e^{-\gamma t} \end{pmatrix}. \tag{3.52}$$

Initial conditions for (3.51a)–(3.51c) are then, respectively,

$$\langle \sigma_-(t)\boldsymbol{s}(t)\rangle = \begin{pmatrix} 0 \\ 1 - e^{-\gamma t} \\ 0 \end{pmatrix}, \tag{3.53a}$$

$$\langle \sigma_+(t)\boldsymbol{s}(t)\rangle = \begin{pmatrix} e^{-\gamma t} \\ 0 \\ 0 \end{pmatrix}, \tag{3.53b}$$

$$\langle \sigma_+(t)\sigma_-(t)\boldsymbol{s}(t)\rangle = \begin{pmatrix} 0 \\ 0 \\ e^{-\gamma t} \end{pmatrix}, \tag{3.53c}$$

where we have used (2.25a) and (2.25b), together with the following:

$$\sigma_+^2 = |2\rangle\langle 1|2\rangle\langle 1| = 0, \tag{3.54a}$$

$$\sigma_-^2 = |1\rangle\langle 2|1\rangle\langle 2| = 0, \tag{3.54b}$$

$$\sigma_+\sigma_-\sigma_+ = |2\rangle\langle 1|1\rangle\langle 2|2\rangle\langle 1| = |2\rangle\langle 1| = \sigma_+, \tag{3.54c}$$

$$\sigma_-\sigma_+\sigma_- = |1\rangle\langle 2|2\rangle\langle 1|1\rangle\langle 2| = |1\rangle\langle 2| = \sigma_-. \tag{3.54d}$$

The nonzero correlation functions obtained from (3.51a)–(3.51c) with initial conditions (3.53) are ($\tau \geq 0$)

$$\langle\sigma_-(t)\sigma_+(t+\tau)\rangle = e^{i\omega_A\tau}e^{-(\gamma/2)\tau}\left(1-e^{-\gamma t}\right), \tag{3.55}$$

$$\langle\sigma_+(t)\sigma_-(t+\tau)\rangle = e^{-i\omega_A\tau}e^{-(\gamma/2)\tau}e^{-\gamma t}, \tag{3.56}$$

$$\langle\sigma_+(t)\sigma_-(t)\sigma_+(t+\tau)\sigma_-(t+\tau)\rangle = e^{-\gamma\tau}e^{-\gamma t}. \tag{3.57}$$

Equation (3.56) provides the result for the emission spectrum. For an ideal detector the probability of detecting a photon of frequency ω during the interval $t = 0$ to $t = T$ is given by [3.3]

$$P(\omega) \propto \int_0^T dt \int_0^T dt'\, e^{-i\omega(t-t')}\langle\sigma_+(t)\sigma_-(t')\rangle. \tag{3.58}$$

We saw how the field at the detector is related to the atomic source operators σ_- and σ_+ in Sect. 2.4 [Eq. (2.61)]. Using (3.56) and

$$\langle\sigma_+(t+\tau)\sigma_-(t)\rangle = \langle\sigma_+(t)\sigma_-(t+\tau)\rangle^*, \tag{3.59}$$

we find, for all t and t',

$$\langle\sigma_+(t)\sigma_-(t')\rangle = e^{i\omega_A(t-t')}e^{-(\gamma/2)(t+t')}. \tag{3.60}$$

Then,

$$\begin{aligned} P(\omega) &\propto \int_0^T dt\, e^{-[(\gamma/2)+i(\omega-\omega_A)]t}\int_0^T dt'\, e^{-[(\gamma/2)-i(\omega-\omega_A)]t'} \\ &\propto \frac{1-e^{-(\gamma/2)T}e^{-i(\omega-\omega_A)T}}{\gamma/2+i(\omega-\omega_A)}\,\frac{1-e^{-(\gamma/2)T}e^{i(\omega-\omega_A)T}}{\gamma/2-i(\omega-\omega_A)}. \end{aligned} \tag{3.61}$$

For long times, $T >> 1/\gamma$, this gives the Lorentzian lineshape

$$P(\omega) \propto \frac{1}{(\gamma/2)^2+(\omega-\omega_A)^2}. \tag{3.62}$$

As a final example we calculate the *spectrum of resonance fluorescence.* This is one of the classic calculations of quantum optics, first performed by Mollow [3.4]. The spectrum is given by

$$S(\omega) = f(\mathbf{r})\frac{1}{2\pi}\int_{-\infty}^{\infty} d\tau\, e^{i\omega\tau}\langle\sigma_+(0)\sigma_-(\tau)\rangle_{ss}, \tag{3.63}$$

where $\langle\sigma_+(0)\sigma_-(\tau)\rangle_{ss} \equiv \lim_{t\to\infty}\langle\sigma_+(t)\sigma_-(t+\tau)\rangle$, and the calculation of the correlation function is to be based on the optical Bloch equations (3.5) (with $\bar{n} = 0$). The function $f(\boldsymbol{r})$ contains the spatial dependence of the dipole radiation resulting from the factor multiplying $\sigma_-(t - r/c)$ in (2.61).

From the solutions (3.9) and (3.10) to the optical Bloch equations we see that, in a rotating frame, the atomic scatterer decays to the steady state

$$\langle\tilde{\sigma}_\mp\rangle_{ss} = e^{\pm i\omega_A t}\langle\sigma_\mp\rangle_{ss} = \pm i\frac{1}{\sqrt{2}}\frac{Y}{1+Y^2}, \tag{3.64a}$$

$$\langle\sigma_z\rangle_{ss} = -\frac{1}{1+Y^2}. \tag{3.64b}$$

However, fluctuations away from this steady state can occur, described by the operators

$$\Delta\tilde{\sigma}_\mp = \tilde{\sigma}_\mp - \langle\tilde{\sigma}_\mp\rangle_{ss}, \tag{3.65a}$$

$$\Delta\sigma_z = \sigma_z - \langle\sigma_z\rangle_{ss}. \tag{3.65b}$$

The fluorescence spectrum therefore decomposes into a coherent component and an incoherent component arising from quantum fluctuations:

$$S(\omega) = S_{coh}(\omega) + S_{inc}(\omega), \tag{3.66}$$

with

$$\begin{aligned} S_{coh}(\omega) &= f(\boldsymbol{r})\frac{1}{2\pi}\int_{-\infty}^{\infty} d\tau\, e^{i(\omega-\omega_A)\tau}\langle\tilde{\sigma}_+\rangle_{ss}\langle\tilde{\sigma}_-\rangle_{ss} \\ &= f(\boldsymbol{r})\frac{1}{2}\frac{Y^2}{(1+Y^2)^2}\,\delta(\omega-\omega_A), \end{aligned} \tag{3.67}$$

and

$$S_{inc}(\omega) = f(\boldsymbol{r})\frac{1}{2\pi}\int_{-\infty}^{\infty} d\tau\, e^{i(\omega-\omega_A)\tau}\langle\Delta\tilde{\sigma}_+(0)\Delta\tilde{\sigma}_-(\tau)\rangle_{ss}. \tag{3.68}$$

Let I_{coh} and I_{inc} denote the coherent and incoherent intensities obtained by integrating (3.67) and (3.68) over all frequencies:

$$I_{coh} = f(\boldsymbol{r})\langle\tilde{\sigma}_+\rangle_{ss}\langle\tilde{\sigma}_-\rangle_{ss} = f(\boldsymbol{r})\frac{1}{2}\frac{Y^2}{(1+Y^2)^2}, \tag{3.69}$$

and

$$I_{inc} = f(\boldsymbol{r})\langle\Delta\tilde{\sigma}_+\Delta\tilde{\sigma}_-\rangle_{ss} = f(\boldsymbol{r})\frac{1}{2}\frac{Y^4}{(1+Y^2)^2}. \tag{3.70}$$

From these we can make some observations about the qualitative form of the spectrum. At weak laser intensities, the ratio $I_{inc}/I_{coh} = Y^2 = 2\Omega^2/\gamma^2$ is very small, and coherent scattering dominates. However, I_{inc}/I_{coh} increases with the laser intensity, and the incoherent spectral component will dominate at high laser intensities. Since the relaxation, or regression, of fluctuations around the steady state will follow a modulated decay similar to that shown by (3.9) and (3.10), we expect this incoherent spectrum to show sidebands at $\omega_A \pm \Omega$.

To calculate the incoherent spectrum we solve for $\langle \Delta\tilde{\sigma}_+(0)\Delta\tilde{\sigma}_-(\tau)\rangle_{ss}$ using the optical Bloch equations and the quantum regression theorem. From (3.5), (3.64), and (3.65),

$$\frac{d}{dt}\langle \Delta\tilde{\sigma}_- \rangle = -i(\Omega/2)\langle \Delta\sigma_z \rangle - \frac{\gamma}{2}\langle \Delta\tilde{\sigma}_- \rangle, \tag{3.71a}$$

$$\frac{d}{dt}\langle \Delta\tilde{\sigma}_+ \rangle = i(\Omega/2)\langle \Delta\sigma_z \rangle - \frac{\gamma}{2}\langle \Delta\tilde{\sigma}_+ \rangle, \tag{3.71b}$$

$$\frac{d}{dt}\langle \Delta\sigma_z \rangle = i\Omega\langle \Delta\tilde{\sigma}_+ \rangle - i\Omega\langle \Delta\tilde{\sigma}_- \rangle - \gamma\Delta\sigma_z, \tag{3.71c}$$

and the quantum regression theorem gives

$$\frac{d}{d\tau}\langle \Delta\tilde{\sigma}_+(0)\Delta\boldsymbol{s}(\tau)\rangle_{ss} = \boldsymbol{M}\langle \Delta\tilde{\sigma}_+(0)\Delta\boldsymbol{s}(\tau)\rangle_{ss}, \tag{3.72}$$

where

$$\Delta\boldsymbol{s} = \begin{pmatrix} \Delta\tilde{\sigma}_- \\ \Delta\tilde{\sigma}_+ \\ \Delta\sigma_z \end{pmatrix}, \tag{3.73}$$

and

$$\boldsymbol{M} = -\frac{\gamma}{2}\begin{pmatrix} 1 & 0 & iY/\sqrt{2} \\ 0 & 1 & -iY/\sqrt{2} \\ i\sqrt{2}Y & -i\sqrt{2}Y & 2 \end{pmatrix}. \tag{3.74}$$

The desired correlation function is the first component of the vector $\langle \Delta\tilde{\sigma}_+(0)$ $\Delta\boldsymbol{s}(\tau)\rangle_{ss}$. The initial conditions are given by

$$\langle \Delta\tilde{\sigma}_+\Delta\boldsymbol{s}\rangle_{ss} = \begin{pmatrix} \frac{1}{2}(1+\langle\sigma_z\rangle_{ss}) - \langle\tilde{\sigma}_+\rangle_{ss}\langle\sigma_-\rangle_{ss} \\ -\langle\tilde{\sigma}_+\rangle_{ss}^2 \\ -\langle\tilde{\sigma}_+\rangle_{ss}(1+\langle\sigma_z\rangle_{ss}) \end{pmatrix}, \tag{3.75}$$

where we have used (2.25), (3.54), and

$$\sigma_+\sigma_z = |2\rangle\langle 1|(|2\rangle\langle 2| - |1\rangle\langle 1|) = -|2\rangle\langle 1| = -\sigma_+, \tag{3.76a}$$

$$\sigma_-\sigma_z = |1\rangle\langle 2|(|2\rangle\langle 2| - |1\rangle\langle 1|) = |1\rangle\langle 2| = \sigma_-. \tag{3.76b}$$

Using the steady-state averages (3.64), we obtain

$$\langle \Delta\tilde{\sigma}_+\Delta\boldsymbol{s}\rangle_{ss} = \frac{1}{2}\frac{Y^2}{(1+Y^2)^2}\begin{pmatrix} Y^2 \\ 1 \\ -iY/\sqrt{2} \end{pmatrix}. \tag{3.77}$$

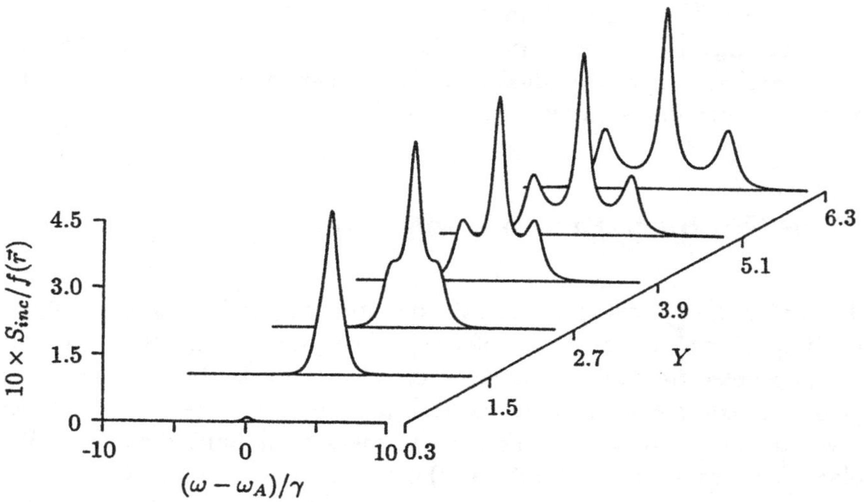

Fig. 3.1. The incoherent fluorescence spectrum as a function of laser intensity.

Equation (3.72) can be solve by finding a matrix $\boldsymbol{S}$ to diagonalize $\boldsymbol{M}$. Multiplying (3.72) on the left by $\boldsymbol{S}$,

$$\frac{d}{d\tau}\boldsymbol{S}\langle\Delta\tilde{\sigma}_+(0)\Delta\boldsymbol{s}(\tau)\rangle_{ss} = (\boldsymbol{S}\boldsymbol{M}\boldsymbol{S}^{-1})\boldsymbol{S}\langle\Delta\tilde{\sigma}_+(0)\Delta\boldsymbol{s}(\tau)\rangle_{ss}, \tag{3.78}$$

and, formally,

$$\langle\Delta\tilde{\sigma}_+(0)\Delta\boldsymbol{s}(\tau)\rangle_{ss} = \boldsymbol{S}^{-1}\exp(\boldsymbol{\Lambda}\tau)\boldsymbol{S}\langle\Delta\tilde{\sigma}_+\Delta\boldsymbol{s}\rangle_{ss}, \tag{3.79}$$

where

$$\boldsymbol{\Lambda} = \boldsymbol{S}\boldsymbol{M}\boldsymbol{S}^{-1} = \mathrm{diag}\left(-\frac{\gamma}{2}, -\frac{3\gamma}{4}+\delta, -\frac{3\gamma}{4}-\delta\right) \tag{3.80}$$

is formed from the eigenvalues of $\boldsymbol{M}$, and the rows (columns) of $\boldsymbol{S}$ ($\boldsymbol{S}^{-1}$) are the left (right) eigenvectors of $\boldsymbol{M}$; δ is defined in (3.12). After some algebra

$$\begin{aligned}
&\langle\Delta\tilde{\sigma}_+(0)\Delta\tilde{\sigma}_-(\tau)\rangle_{ss} \\
&\quad= \frac{1}{4}\frac{Y^2}{1+Y^2}e^{-(\gamma/2)\tau} \\
&\qquad- \frac{1}{8}\frac{Y^2}{(1+Y^2)^2}\left[1-Y^2+(1-5Y^2)\frac{(\gamma/4)}{\delta}\right]e^{-[(3\gamma/4)-\delta]\tau} \\
&\qquad- \frac{1}{8}\frac{Y^2}{(1+Y^2)^2}\left[1-Y^2-(1-5Y^2)\frac{(\gamma/4)}{\delta}\right]e^{-[(3\gamma/4)+\delta]\tau}.
\end{aligned} \tag{3.81}$$

Expressions for the incoherent spectrum are calculated from (3.68) and (3.81); clearly these involve a sum of three Lorentzian components. It is easy to see that in the strong-field limit, $Y^2 >> 1$ $(\Omega^2 >> \gamma^2)$, where incoherent scattering dominates, this calculation gives the well-known Mollow, or Stark, triplet. Figure 3.1 illustrates the development of the incoherent fluorescence spectrum with laser intensity.

3.4 The Hanbury-Brown-Twiss effect

In addition to the optical spectrum, the quantum regression theorem allows us to analyze various properties of the source photon statistics. We will discuss photoelectric counting in some detail in a future lecture, so let us postpone any comment on the connection between the correlation function we now calculate and the scheme used to measure it until that time. We calculate the second-order (in intensity) correlation function, and note only that this quantity is proportional to the joint probability for detecting two photons in short counting intervals centered at two different times.

The second-order correlation function is given by a normal-ordered, time-ordered average; thus, there is no free field contribution if the reservoir is in the vacuum state (Sects. 1.4 and 1.5). For a cavity mode driven by thermal light the second-order correlation function of the output field is given by $\langle a^\dagger(t)a^\dagger(t+\tau)a(t+\tau)a(t)\rangle = \langle a^\dagger(t)\hat{n}(t+\tau)a(t)\rangle$ if we refer to the light radiated through a mirror that is not illuminated by the thermal light. To calculate this correlation function we first write (3.2) in the form

$$\frac{d}{dt}\begin{pmatrix}\langle\hat{n}\rangle\\ \bar{n}\end{pmatrix} = \begin{pmatrix}-2\kappa & 2\kappa\\ 0 & 0\end{pmatrix}\begin{pmatrix}\langle\hat{n}\rangle\\ \bar{n}\end{pmatrix}. \tag{3.82}$$

We then set $\hat{A}_1 = \hat{n} = a^\dagger a$ and $\hat{A}_2 = \bar{n}$ (a constant), and from (3.37) and (3.41), with $\hat{O}_1 = a^\dagger$ and $\hat{O}_2 = a$,

$$\frac{d}{d\tau}\begin{pmatrix}\langle a^\dagger(t)\hat{n}(t+\tau)a(t)\rangle\\ \bar{n}\langle\hat{n}(t)\rangle\end{pmatrix} = \begin{pmatrix}-2\kappa & 2\kappa\\ 0 & 0\end{pmatrix}\begin{pmatrix}\langle a^\dagger(t)\hat{n}(t+\tau)a(t)\rangle\\ \bar{n}\langle\hat{n}(t)\rangle\end{pmatrix}. \tag{3.83}$$

Thus,

$$\langle a^\dagger(t)\hat{n}(t+\tau)a(t)\rangle = \langle a^\dagger(t)\hat{n}(t)a(t)\rangle e^{-2\kappa\tau} + \bar{n}\langle\hat{n}(t)\rangle(1-e^{-2\kappa\tau}). \tag{3.84}$$

We obtained an expression for $\langle\hat{n}(t)\rangle$ in (3.3). The calculation of $\langle a^\dagger(t)\hat{n}(t)$ $a(t)\rangle$ follows similar lines and gives

$$\begin{aligned}\langle a^\dagger(t)\hat{n}(t)a(t)\rangle = {}& \left[\langle\hat{n}^2(0)\rangle - \langle\hat{n}(0)\rangle\right]e^{-4\kappa t} + 2\bar{n}(1-e^{-2\kappa t})\\ &\times\left[2\langle\hat{n}(0)\rangle e^{-2\kappa t} + \bar{n}(1-e^{-2\kappa t})\right].\end{aligned} \tag{3.85}$$

Now, substituting (3.3) and (3.85) into (3.84),

$$
\begin{aligned}
&\langle a^\dagger(t)a^\dagger(t+\tau)a(t+\tau)a(t)\rangle \\
&= \{[(\langle\hat{n}^2(0)\rangle - \langle\hat{n}(0)\rangle]e^{-4\kappa t} + 2\bar{n}(1-e^{-2\kappa t})[2\langle\hat{n}(0)\rangle e^{-2\kappa t} \\
&\quad +\bar{n}(1-e^{-2\kappa t})]\}e^{-2\kappa\tau} + \bar{n}\left[\langle\hat{n}(0)\rangle e^{-2\kappa t} + \bar{n}(1-e^{-2\kappa t})\right](1-e^{-2\kappa\tau}).
\end{aligned}
\tag{3.86}
$$

In the long-time limit,

$$
\begin{aligned}
\langle a^\dagger(0)a^\dagger(\tau)a(\tau)a(0)\rangle_{ss} &\equiv \lim_{t\to\infty}\langle a^\dagger(t)a^\dagger(t+\tau)a(t+\tau)a(t)\rangle \\
&= \bar{n}^2(1+e^{-2\kappa\tau}).
\end{aligned}
\tag{3.87}
$$

This expression describes the well-known Hanbury-Brown-Twiss effect, or photon bunching, for thermal light [3.5]; at zero delay the correlation function has twice its value for long delays ($2\kappa\tau \gg 1$).

3.5 Photon antibunching

Based on its spectrum alone, for weak laser intensities atomic fluorescence is coherent – the fluorescence field shows first-order coherence. But is it coherent to higher orders? In the long-time limit, second-order coherence requires that

$$
G_{ss}^{(2)}(\tau) = f(\boldsymbol{r})^2\langle\sigma_+(0)\sigma_+(\tau)\sigma_-(\tau)\sigma_-(0)\rangle_{ss} = \left[f(\boldsymbol{r})\langle\sigma_+\rangle_{ss}\langle\sigma_-\rangle_{ss}\right]^2.
$$

Clearly this is never satisfied for $\tau = 0$, since $\langle\sigma_+\rangle_{ss}^2$ and $\langle\sigma_-\rangle_{ss}^2$ are not zero [from (3.64a)], but σ_+^2 and σ_-^2 vanish identically. The latter simply states that a two-state atom cannot be sequentially raised or lowered twice; two photons cannot be absorbed or emitted simultaneously. The detection of one photon sets the atom in its ground state, and a second photon cannot be detected until the atom has been reexcited. We might predict, then, that the probability for detecting two photons is just the probability for detecting the first photon, multiplied by the probability for detecting a second photon at the time $t = \tau$, given that the atom was in its ground state at $t = 0$. We are suggesting that

$$
G_{ss}^{(2)}(\tau) = f(\boldsymbol{r})^2\langle 2|\rho_{ss}|2\rangle\langle 2|\rho(\tau)|2\rangle_{\rho(0)=|1\rangle\langle 1|}.
\tag{3.88}
$$

This is clearly zero for $\tau = 0$, and gives independent detection events for large τ, as $\rho(\tau) \to \rho_{ss}$. We will use the quantum regression theorem to prove this result. The result $G^{(2)}(0) = 0$ is impossible for any classical field; it implies photon antibunching instead of the photon bunching of the Hanbury-Brown-Twiss effect. Photon antibunching in resonance fluorescence is important as the first experimentally tested example of what are now referred to as nonclassical properties of a photoemissive source [3.6, 3.7].

To prove (3.88), first let us consider the formal solution to the optical Bloch equations. In a rotating frame, (3.5a)–(3.5c) can be written in the vector form

$$\langle \dot{\boldsymbol{s}} \rangle = \boldsymbol{M} \langle \boldsymbol{s} \rangle + \boldsymbol{b}, \tag{3.89}$$

where

$$\boldsymbol{s} = \begin{pmatrix} \tilde{\sigma}_- \\ \tilde{\sigma}_+ \\ \sigma_z \end{pmatrix}, \tag{3.90}$$

$\boldsymbol{M}$ is the 3×3 matrix given by (3.74), and

$$\boldsymbol{b} = -\gamma \begin{pmatrix} 0 \\ 0 \\ 1 \end{pmatrix}. \tag{3.91}$$

Then

$$\frac{d}{dt}\left(\langle \boldsymbol{s} \rangle + \boldsymbol{M}^{-1}\boldsymbol{b}\right) = \boldsymbol{M}\left(\langle \boldsymbol{s} \rangle + \boldsymbol{M}^{-1}\boldsymbol{b}\right), \tag{3.92}$$

and

$$\langle \boldsymbol{s}(t) \rangle = -\boldsymbol{M}^{-1}\boldsymbol{b} + \exp(\boldsymbol{M}t)\left(\langle \boldsymbol{s}(0) \rangle + \boldsymbol{M}^{-1}\boldsymbol{b}\right). \tag{3.93}$$

Now

$$\begin{aligned} G_{ss}^{(2)}(\tau) &= f(\boldsymbol{r})^2 \langle \sigma_+(0)\sigma_+(\tau)\sigma_-(\tau)\sigma_-(0)\rangle_{ss} \\ &= f(\boldsymbol{r})^2\, \tfrac{1}{2}\left[\langle \sigma_+\sigma_-\rangle_{ss} + \langle \sigma_+(0)\sigma_z(\tau)\sigma_-(0)\rangle_{ss}\right], \end{aligned} \tag{3.94}$$

where we have used (2.25a). We can calculate the correlation function $\langle \sigma_+(0)\sigma_z(\tau)\sigma_-(0)\rangle_{ss}$ using the quantum regression theorem, as the third component of the vector $\langle \sigma_+ \boldsymbol{s}(\tau)\sigma_-\rangle_{ss}$. To find the equation of motion for this vector, the quantum regression theorem tells us to remove the angular brackets from (3.89) ($\boldsymbol{b}$ is a constant vector multiplied by the expectation of the identity operator), multiply on the left by $\sigma_+(0)$ and on the right by $\sigma_-(0)$, and replace the angular brackets; thus

$$\frac{d}{d\tau}\langle \sigma_+(0)\boldsymbol{s}(\tau)\sigma_-(0)\rangle_{ss} = \boldsymbol{M}\left[\langle \sigma_+(0)\boldsymbol{s}(\tau)\sigma_-(0)\rangle_{ss} + \langle \sigma_+\sigma_-\rangle_{ss}\boldsymbol{M}^{-1}\boldsymbol{b}\right]. \tag{3.95}$$

The formal solution to this equation is

$$\begin{aligned} \langle \sigma_+(0)\boldsymbol{s}(\tau)\sigma_-(0)\rangle_{ss} = &-\langle \sigma_+\sigma_-\rangle_{ss}\boldsymbol{M}^{-1}\boldsymbol{b} \\ &+ \exp(\boldsymbol{M}\tau)\left[\langle \sigma_+\boldsymbol{s}\,\sigma_-\rangle_{ss} + \langle \sigma_+\sigma_-\rangle_{ss}\boldsymbol{M}^{-1}\boldsymbol{b}\right], \end{aligned} \tag{3.96}$$

with initial conditions

$$\langle\sigma_+ \boldsymbol{s}\,\sigma_-\rangle_{ss} = \langle\sigma_+\sigma_-\rangle_{ss}\begin{pmatrix}0\\0\\-1\end{pmatrix}, \tag{3.97}$$

where we have used results (3.54) and (3.76). Then (3.96), (3.97), and (3.93) give

$$\begin{aligned}&\langle\sigma_+(0)\boldsymbol{s}(\tau)\sigma_-(0)\rangle_{ss}\\ &\quad= \langle\sigma_+\sigma_-\rangle_{ss}\left\{-\boldsymbol{M}^{-1}\boldsymbol{b} + \exp(\boldsymbol{M}\tau)\left[\begin{pmatrix}0\\0\\-1\end{pmatrix} + \boldsymbol{M}^{-1}\boldsymbol{b}\right]\right\}\\ &\quad= \langle\sigma_+\sigma_-\rangle_{ss}\langle\boldsymbol{s}(\tau)\rangle_{\rho(0)=|1\rangle\langle 1|}.\end{aligned} \tag{3.98}$$

Here, we noted that $\begin{pmatrix}0\\0\\-1\end{pmatrix}$ is simply the initial condition $\langle\boldsymbol{s}(0)\rangle$ for an atom prepared in its ground state – i.e. with $\rho(0) = |1\rangle\langle 1|$. Substituting the third component of (3.98) into (3.94) establishes our result:

$$\begin{aligned}G_{ss}^{(2)}(\tau) &= f(\boldsymbol{r})^2\langle\sigma_+\sigma_-\rangle_{ss}\,\tfrac{1}{2}\Big(1 + \langle\sigma_z(\tau)\rangle_{\rho(0)=|1\rangle\langle 1|}\Big)\\ &= f(\boldsymbol{r})^2\langle 2|\rho_{ss}|2\rangle\langle 2|\rho(\tau)|2\rangle_{\rho(0)=|1\rangle\langle 1|}.\end{aligned} \tag{3.99}$$

Note that this calculation is independent of the form of $\boldsymbol{M}$. Thus, while (3.74) only gives $\boldsymbol{M}$ for perfect resonance, (3.99) also holds off resonance.

The factorized form of (3.99) actually follows very simply, and quite generally, from the quantum regression theorem in the form (3.34):

$$\begin{aligned}G_{ss}^{(2)}(\tau) &= f(\boldsymbol{r})^2\langle\sigma_+(0)\sigma_+(\tau)\sigma_-(\tau)\sigma_-(0)\rangle_{ss}\\ &= f(\boldsymbol{r})^2\operatorname{tr}\left\{e^{\mathcal{L}\tau}\left[\sigma_-(0)\rho_{ss}\sigma_+(0)\right]\sigma_+(0)\sigma_-(0)\right\}\\ &= f(\boldsymbol{r})^2\operatorname{tr}\left\{e^{\mathcal{L}\tau}\left[|1\rangle\langle 2|\rho_{ss}|2\rangle\langle 1|\right]|2\rangle\langle 2|\right\}\\ &= f(\boldsymbol{r})^2\langle 2|\rho_{ss}|2\rangle\langle 2|e^{\mathcal{L}\tau}\big(|1\rangle\langle 1|\big)|2\rangle,\end{aligned}$$

and $\langle 2|e^{\mathcal{L}\tau}\big(|1\rangle\langle 1|\big)|2\rangle$ is just a formal expression for $\langle 2|\rho(\tau)|2\rangle_{\rho(0)=|1\rangle\langle 1|}$.

Equation (3.10) provides the solution for $\langle\sigma_z(t)\rangle_{\rho(0)=|1\rangle\langle 1|}$ from which the explicit form for $G_{ss}^{(2)}(\tau)$ may be written down. We normalize $G_{ss}^{(2)}(\tau)$ by its factorized form for independent photon detection in the large-delay limit and write

$$\begin{aligned}g_{ss}^{(2)}(\tau) &= \left[\lim_{\tau\to\infty} G_{ss}^{(2)}(\tau)\right]^{-1} G_{ss}^{(2)}(\tau)\\ &= 1 - e^{-(3\gamma/4)\tau}\left(\cosh\delta\tau + \frac{3\gamma/4}{\delta}\sinh\delta\tau\right).\end{aligned} \tag{3.100}$$

This expression is plotted in Fig. 3.2. For a field possessing second-order coherence $g_{ss}^{(2)}(\tau) = 1$; the two photons are detected independently for all decay times. In this case a detector responds to the incident light with a completely random sequence of photopulses. This provides reference against which the "antibunching" of photopulses is defined. All of the curves in Fig. (3.2) show photon antibunching because $g_{ss}^{(2)}(0)$ falls below unity, the value for independent photocounts. We will discuss the reasons for this being nonclassical when we come to the treatment of photoelectric detection and photon counting.

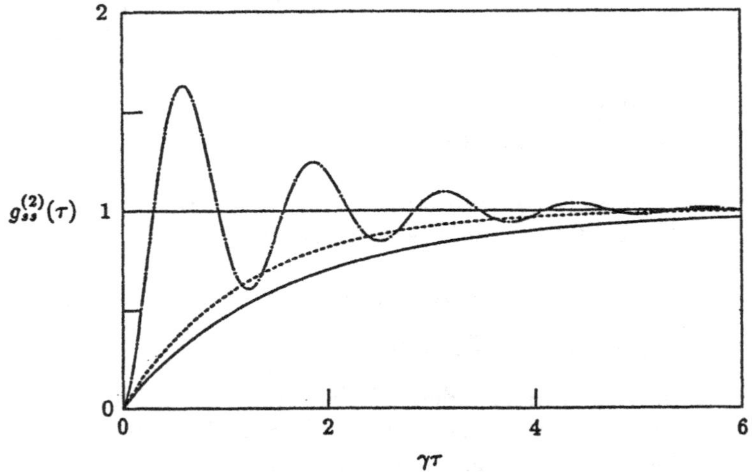

Fig. 3.2. The normalized second-order correlation function (3.100): (a) (solid curve) $8Y^2 = 0.01 \ll 1$ ($\delta \approx \gamma/4$); (b) (dashed curve) $8Y^2 = 1$ ($\delta = 0$); (c) (dot-dash curve) $8Y^2 = 400 \gg 1$ ($\delta \approx i\Omega$).

References

[3.1] M. Lax, *Phys. Rev.* **129**, 2342 (1963).
[3.2] M. Lax, *Phys. Rev.* **157**, 213 (1967).
[3.3] R. J. Glauber, "Optical Coherence and Photon Statistics," in *Quantum Optics and Electronics*, ed. by C. DeWitt, A. Blandin, and C. Cohen-Tannoudji, Gordon and Breach: London, 1965, pp. 78ff – in particular, consider Eq. (4.11) with a sharply peaked (δ-function) sensitivity function $s(\omega)$.
[3.4] B. R. Mollow, *Phys. Rev.* **188**, 1969 (1969).
[3.5] R. Hanbury-Brown and R. Q. Twiss, *Nature* **177**, 27 (1956); **178**, 1046 (1956); *Proc. R. Soc. Lond.* A **242**, 300 (1957); **243**, 291 (1957).
[3.6] H. J. Carmichael and D. F. Walls, *J. Phys.* B **9**, L43 (1976); *ibid*, 1199 (1976).

[3.7] H. J. Kimble, M. Dagenais, and L. Mandel, *Phys. Rev. Lett.* **39**, 691 (1977).

Lecture 4 – Standard Methods of Analysis II

It is generally not possible to solve an operator master equation directly to find $\rho(t)$ in operator form. We have seen, however, that alternative methods of analysis are available. We can derive equations of motion for expectation values and solve these for time-dependent operator averages. Alternatively, we may choose a representation and take matrix elements of the master equation to obtain equations of motion for the matrix elements of ρ. We have also seen how equations of motion for one-time operator averages can be used to obtain equations of motion for two-time averages (correlation functions) using the quantum regression theorem.

We are now going to meet an entirely new approach to the problem of solving the operator master equation and calculating operator averages and correlation functions. We will only explicitly consider master equations involving electromagnetic field modes, but the methods we discuss can be generalized to systems that involve two-state atoms.

4.1 Quantum-Classical Correspondence

The new approach sets up a correspondence between quantum-mechanical operators and ordinary (classical) functions, such that quantities of interest in a quantum-mechanical problem can be calculated using the methods of classical statistical physics. Under this correspondence the operator master equation transforms into a partial differential equation for a quasidistribution function that corresponds to (represents) ρ. Operator averages, written in an appropriate order (e.g. normal order), are calculated by integrating functions of classical phase-space variables against the quasidistribution function, in the same manner in which we take classical phase-space averages. This quantum-classical correspondence is particularly appealing when the partial differential equation corresponding to the operator master equation is a Fokker-Planck equation. Fokker-Planck equations are familiar from classical statistical physics and have been studied extensively [4.1]. When the operator master equation becomes a Fokker-Planck equation, analogies can be drawn between classical fluctuation phenomena and fluctuations generated by the quantum dynamics. This helps us develop an intuition for the effects of quantum fluctuations. Also, mathematical techniques that were developed for analyzing Fokker-Plank equations in their traditional setting can be sequestered to help solve a quantum-mechanical problem.

There are, in fact, many ways in which to set up a quantum-classical correspondence. We will mention only three. The original ideas go back to the work of Wigner [4.2]. However, Wigner was interested in general questions of quantum statistical mechanics, not specifically in quantum-optical applications; wide use of the methods of quantum-classical correspondence for problems in quantum optics began with the work of Glauber [4.3] and Sudarshan [4.4]. These authors independently developed what is now commonly known as the Glauber-Sudarshan P representation, or simply the P representation, for the electromagnetic field. This representation is based upon a correspondence in which normal-ordered operator averages are calculated as classical phase-space averages; it has been tailored for the special role played by normal-ordered averages in the theory of photodetection and quantum coherence [4.3, 4.5, 4.6]. The Wigner representation gives the averages of operators in the Weyl, or symmetric, ordering.

The Glauber-Sudarshan P representation was introduced primarily for the description of statistical mixtures of coherent states – the closest approach within the quantum theory to the states of the electromagnetic field described by the classical statistical theory of optics. An understanding of this representation can therefore be built on a few simple properties of the coherent states. Formal definition of the P representation can, alternatively, be given without any mention of the coherent states; this is the more useful approach when we want to generalize the methods to other representations for the field, and to representations for collections of two-state atoms. We will look at both definitions of the P representation. We begin with the definition in terms of the coherent states.

The coherent state $|\alpha\rangle$ is the right eigenstate of the annihilation operator a with complex eigenvalue α:

$$a|\alpha\rangle = \alpha|\alpha\rangle, \qquad \langle\alpha|a^\dagger = (a|\alpha\rangle)^\dagger = \alpha^*\langle\alpha|. \tag{4.1}$$

The Glauber-Sudarshan P representation relies on the fact that the coherent states are not orthogonal. In technical terms they then form an overcomplete basis, and, as a consequence, it is often possible to expand ρ as a diagonal sum over coherent states:

$$\rho = \int d^2\alpha\, |\alpha\rangle\langle\alpha|P(\alpha). \tag{4.2}$$

This representation for ρ is appealing because the function $P(\alpha)$ plays a role which is rather analogous to a classical probability distribution. For the expectation values of operators written in normal order (creation operators to the left and annihilation operators to the right), on substituting the expansion (4.2) for ρ, we obtain

$$\begin{aligned}\langle a^{\dagger p} a^q \rangle &= \mathrm{tr}\left(\rho a^{\dagger p} a^q\right) \\ &= \mathrm{tr}\left(\left(\int d^2\alpha\, |\alpha\rangle\langle\alpha| P(\alpha) a^{\dagger p} a^q\right)\right. \\ &= \int d^2\alpha\, P(\alpha) \alpha^{*p} \alpha^q. \end{aligned} \tag{4.3}$$

Normal-ordered averages are therefore calculated in the same way that averages are calculated in classical statistics, with $P(\alpha)$ playing the role of the probability distribution. Setting $p = q = 0$ we find that the integral of $P(\alpha)$ over the complex plane is given by $\mathrm{tr}(\rho) = 1$; thus, $P(\alpha)$ is normalized like a classical probability distribution.

The analogy between $P(\alpha)$ and a classical distribution must be made with reservation, however. In the Fock-state representation $\rho_{n,n} = \langle n|\rho|n\rangle$ is an actual probability; it is the probability that the cavity mode will be found to contain n photons. But, because of the orthogonality of the Fock states it is not possible to expand an arbitrary ρ in terms of the diagonal matrix elements $\rho_{n,n}$ alone. The coherent states are not orthogonal, and it is therefore possible to make a diagonal expansion for ρ without automatically requiring that the off-diagonal coherent state matrix elements vanish. However, along with this greater versatility we must now accept that $P(\alpha)$ is not strictly a probability. From (4.2), the nonorthogonality of the coherent states gives

$$\langle\alpha|\rho|\alpha\rangle = \int d^2\lambda\, e^{-|\lambda-\alpha|^2} P(\lambda), \tag{4.4}$$

where we have used $|\langle\alpha|\lambda\rangle|^2 = e^{-|\lambda-\alpha|^2}$. Since $e^{-|\lambda-\alpha|^2}$ is not a δ-function, $\langle\alpha|\rho|\alpha\rangle \neq P(\alpha)$; only when $P(\lambda)$ is sufficiently broad compared to the Gaussian filter in (4.4) does it approximate a probability. Also, although the probability $\langle\alpha|\rho|\alpha\rangle$ must be positive, (4.4) does not require $P(\alpha)$ to be so. Thus, unlike a classical probability, $P(\alpha)$ can take negative values over a limited range. $P(\alpha)$ is not, therefore, a probability distribution, and it is often referred to as a *quasidistribution function*. We will simply use the word "distribution".

It is clear from (4.2) that the coherent state $|\alpha_0\rangle$ – density operator $\rho = |\alpha_0\rangle\langle\alpha_0|$ – is represented by the distribution

$$P(\alpha) = \delta^{(2)}(\alpha - \alpha_0) = \delta(x - x_0)\delta(y - y_0), \tag{4.5}$$

where $\alpha = x + iy$ and $\alpha_0 = x_0 + iy_0$. Now the obvious question is, can we find a diagonal representation for any density operator? To answer this question we must try to invert (4.2). This is made possible using the relationship

$$\begin{aligned}\mathrm{tr}\left(\rho e^{iz^* a^\dagger} e^{iza}\right) &= \mathrm{tr}\left\{\left[\int d^2\alpha\, |\alpha\rangle\langle\alpha| P(\alpha)\right] e^{iz^* a^\dagger} e^{iza}\right\} \\ &= \int d^2\alpha\, P(\alpha) e^{iz^*\alpha^*} e^{iz\alpha}. \end{aligned} \tag{4.6}$$

Equation (4.6) is just a two-dimensional Fourier transform. The inverse transform gives

$$P(\alpha) = \frac{1}{\pi^2}\int d^2z \,\mathrm{tr}\left(\rho e^{iz^* a^\dagger} e^{iza}\right) e^{-iz^*\alpha^*} e^{-iz\alpha}. \tag{4.7}$$

If the Fourier transform of the function defined by the trace in (4.7) exists for a given density operator ρ, we have our P distribution representing that density operator. A general expression for $P(\alpha)$ in terms of the Fock-state representation of ρ can be obtained from (4.7) in the form

$$P(\alpha) = \frac{1}{\pi^2}\int d^2z \left(\sum_{n=0}^{\infty}\sum_{m=0}^{\infty}\sum_{k=0}^{\infty} \rho_{n+k,m+k}\frac{\sqrt{(n+k)!}\sqrt{(m+k)!}}{k!} \times \frac{(iz^*)^m}{m!}\frac{(iz)^n}{n!}\right) e^{-iz^*\alpha^*} e^{-iz\alpha}. \tag{4.8}$$

Substituting $\rho = |\alpha_0\rangle\langle\alpha_0|$ into (4.7) and the Fock-state representation for the coherent state into (4.8) we find that both of these equations reproduce the P distribution for the coherent state given by (4.5). For a thermal state [the one mode version of (1.22)] (4.8) leads to the distribution

$$P(\alpha) = \frac{1}{\pi\bar{n}}\exp\left(-\frac{|\alpha|^2}{\bar{n}}\right), \tag{4.9}$$

where $\bar{n}$ is given by (1.31). Now, consider the P distribution representing a Fock state. We will take $\rho = |l\rangle\langle l|$, where l can be any non-negative integer. From (4.8), we have

$$P(\alpha) = \frac{1}{\pi^2}\int d^2z \left(\sum_{k=0}^{l}\frac{(-1)^k|z|^{2k}}{k!}\frac{l!}{k!(l-k)!}\right) e^{-iz^*\alpha^*} e^{-iz\alpha}. \tag{4.10}$$

Here there is a problem. Since the summation in (4.10) does not extend to infinity, the expression inside the bracket is a polynomial, and it clearly diverges for $|z| \to \infty$. Thus, this Fourier transform does not exist in the ordinary sense; it would appear that we cannot represent a Fock state using only a diagonal expansion in coherent states. However, there is a way out of this difficulty. If we write

$$\delta^{(2)}(\alpha) = \frac{1}{\pi^2}\int d^2z \, e^{-iz^*\alpha^*} e^{-iz\alpha}, \tag{4.11}$$

and use the ordinary rules of differentiation inside the integral, we may write (4.10) as

$$P(\alpha) = \sum_{k=0}^{l}\frac{l!}{k!(l-k)!}\frac{1}{k!}\frac{\partial^{2k}}{\partial\alpha^k\partial\alpha^{*k}}\delta^{(2)}(\alpha), \tag{4.12}$$

where we take derivatives with respect to complex conjugate variables by reading the complex variable and its conjugate as two independent quantities. In (4.12) the Fock state is given a P representation in terms of a generalized function – a "distribution" in the technical sense of Schwartz distributions and tempered distributions [4.7-4.9].

In general, then, the P representation requires that a density operator be represented by a generalized function. If generalized functions are used any state of the quantized cavity mode may be given a diagonal representation [4.10]. But applications of the P representation in quantum optics have largely been limited to situations in which $P(\alpha)$ exists as an ordinary function, as it does, for example, for a thermal state [Eq. (4.9)]. As stated earlier, our main objective for introducing the quantum-classical correspondence is to cast the quantum-mechanical theory into a form closely analogous to a classical statistical theory. $P(\alpha)$ is never strictly a probability for observing the coherent state $|\alpha\rangle$, but it can take the *form* of a probability distribution, and when it does, this can be used to aid our intuition – for example, the phase-independent distribution given by (4.9) agrees with our classical picture of a field mode subject to thermal fluctuations.

We now look at the alternative way of defining the P representation. This second approach leaves the relationship to coherent states somewhat hidden, but introduces a method which can readily be generalized – to define representations based on different operator orderings, and to define representations for collections of two-state atoms. We have just met two relationships which might suggest the new approach to us. In (4.6) and (4.7) we saw that the Fourier transform of $P(\alpha)$ played an important role. Why not begin from the function appearing on the left-hand side of (4.6) and *define* $P(\alpha)$ to be its Fourier transform. Indeed, this approach is suggested on the more general grounds that the function

$$\chi_N(z,z^*) = \mathrm{tr}\left(\rho e^{iz^*a^\dagger} e^{iza}\right) \tag{4.13}$$

that appears on the left-hand side of (4.6) is a *characteristic function* in the usual sense of statistical physics [4.11]; it determines all *normal-ordered operator averages* via the prescription

$$\begin{aligned}\langle a^{\dagger p} a^q\rangle &= \mathrm{tr}\left(\rho a^{\dagger p} a^q\right)\\ &= \frac{\partial^{p+q}}{\partial(iz^*)^p \partial(iz)^q}\,\chi_N(z,z^*)\bigg|_{z=z^*=0}\end{aligned} \tag{4.14}$$

The definition of a distribution for calculating normal-ordered averages follows quite naturally from this result. If we define $P(\alpha,\alpha^*)$ to be the two-dimensional Fourier transform of $\chi_N(z,z^*)$:

$$P(\alpha,\alpha^*) = \frac{1}{\pi^2}\int d^2z\,\chi_N(z,z^*)e^{-iz^*\alpha^*}e^{-iz\alpha}, \tag{4.15}$$

with the inverse relationship

$$\chi_N(z,z^*) = \int d^2\alpha\, P(\alpha,\alpha^*) e^{iz^*\alpha^*} e^{iz\alpha}, \tag{4.16}$$

then, from (4.14) and (4.16),

$$\begin{aligned}\langle a^{\dagger p} a^q \rangle &= \frac{\partial^{p+q}}{\partial(iz^*)^p \partial(iz)^q} \int d^2\alpha\, P(\alpha,\alpha^*) e^{iz^*\alpha^*} e^{iz\alpha} \Big|_{z=z^*=0} \\ &= \int d^2\alpha\, P(\alpha,\alpha^*) \alpha^{*p} \alpha^q. \end{aligned} \tag{4.17}$$

Equation (4.16) is just (4.6), and (4.17) reproduces (4.3). [Note that it is convenient now to read P as a function of the two independent variables α and α^*.]

Many variations on the scheme outlined in (4.13)–(4.17) can be devised. We mention just two. First, if we wish to calculate antinormal-ordered averages, the rather obvious generalization of (4.13) is to introduce

$$\chi_A(z,z^*) = \mathrm{tr}\left(\rho e^{iza} e^{iz^* a^\dagger}\right), \tag{4.18}$$

and define the distribution $Q(\alpha,\alpha^*)$ as the Fourier transform of $\chi_A(z,z^*)$:

$$Q(\alpha,\alpha^*) = \frac{1}{\pi^2} \int d^2z\, \chi_A(z,z^*) e^{-iz^*\alpha^*} e^{-iz\alpha}. \tag{4.19}$$

Then, in place of (4.14), *antinormal-ordered operator averages* are given by

$$\langle a^q a^{\dagger p} \rangle = \int d^2\alpha\, Q(\alpha,\alpha^*) \alpha^{*p} \alpha^q. \tag{4.20}$$

The representation based on the distribution $Q(\alpha,\alpha^*)$ is known as the Q *representation*. It also has a simple relationship to the coherent states. Consider (4.19) with $\chi_A(z,z^*)$ substituted explicitly from (4.18) and the unit operator judiciously introduced from the completeness relation for the coherent states. We find

$$\begin{aligned} Q(\alpha,\alpha^*) &= \frac{1}{\pi^2} \int d^2z\, \mathrm{tr}\left[\rho e^{iza} \left(\frac{1}{\pi} \int d^2\lambda\, |\lambda\rangle\langle\lambda|\right) e^{iz^* a^\dagger}\right] e^{-iz^*\alpha^*} e^{-iz\alpha} \\ &= \frac{1}{\pi} \int d^2\lambda\, \langle\lambda|\rho|\lambda\rangle \left[\frac{1}{\pi^2} \int d^2z\, e^{iz^*(\lambda^*-\alpha^*)} e^{iz(\lambda-\alpha)}\right] \\ &= \frac{1}{\pi} \int d^2\lambda\, \langle\lambda|\rho|\lambda\rangle \delta^{(2)}(\lambda-\alpha) \\ &= \frac{1}{\pi} \langle\alpha|\rho|\alpha\rangle. \end{aligned} \tag{4.21}$$

Thus, $\pi Q(\alpha,\alpha^*)$ is the diagonal matrix element of the density operator taken with respect to the coherent state $|\alpha\rangle$. It is therefore strictly a probability – the probability for observing the coherent state $|\alpha\rangle$.

Finally, we consider the originator of them all, the Wigner representation. The *Wigner representation* is defined by introducing a third characteristic function:

$$\chi_S(z, z^*) = \mathrm{tr}\left(\rho e^{iz^* a^\dagger + iza}\right). \tag{4.22}$$

The Wigner distribution $W(\alpha, \alpha^*)$ is the Fourier transform of $\chi_S(z, z^*)$:

$$W(\alpha, \alpha^*) = \frac{1}{\pi^2}\int d^2z\, \chi_S(z, z^*) e^{-iz^*\alpha^*} e^{-iz\alpha}. \tag{4.23}$$

The relationship between the Wigner distribution and operator averages is a little more complicated than the relationship between the P and Q distributions and operator averages. In terms of position and momentum variables (proportional to the real and imaginary parts of α) the moments of $W(\alpha, \alpha^*)$ give the averages of operators placed in Weyl order [4.12]. The relevant quantities for quantum optics are operator averages corresponding to moments of the complex variables α and α^*. These are the *symmetric-ordered operator averages*; we have

$$\langle (a^{\dagger p} a^q)_S \rangle = \int d^2\alpha\, W(\alpha, \alpha^*) \alpha^{*p} \alpha^q, \tag{4.24}$$

where $(a^{\dagger p} a^q)_S$ denotes the average of $(p+q)!/(p!q!)$ possible orderings of p creation operators and q annihilation operators – for example:

$$(a^\dagger a)_S = \tfrac{1}{2}(a^\dagger a + a a^\dagger), \tag{4.25a}$$

$$(a^{\dagger 2} a)_S = \tfrac{1}{3}(a^{\dagger 2} a + a^\dagger a a^\dagger + a a^{\dagger 2}), \tag{4.25b}$$

$$(a^\dagger a^2)_S = \tfrac{1}{3}(a^\dagger a^2 + a a^\dagger a + a^2 a^\dagger). \tag{4.25c}$$

$$\vdots$$

4.2 Fokker-Planck equation for a cavity mode driven by thermal light

The usefulness of the quantum-classical correspondence lies not so much in its ability to provide a representation for ρ, but in the fact that it often allows the master equation to be converted into a Fokker-Planck equation. Let us see how this works for the master equation (1.47). We will perform the calculation in the P representation, and then note at the end how the Fokker-Planck equation is changed if either the Q or Wigner representation is used.

We first derive an equation of motion for the characteristic function. From the definition of χ_N,

$$\frac{\partial \chi_N}{\partial t} = \frac{\partial}{\partial t}\mathrm{tr}\left(\rho\, e^{iz^* a^\dagger} e^{iza}\right) = \mathrm{tr}\left(\dot{\rho}\, e^{iz^* a^\dagger} e^{iza}\right). \tag{4.26}$$

Then, the master equation (1.47) gives

$$\begin{aligned}\frac{\partial \chi_N}{\partial t} = \mathrm{tr}\Big\{\big[&-i\omega_C(a^\dagger a\rho - \rho a^\dagger a) + \kappa(2a\rho a^\dagger - a^\dagger a\rho - \rho a^\dagger a)\\ &+2\kappa\bar{n}(a\rho a^\dagger + a^\dagger \rho a - a^\dagger a\rho - \rho a a^\dagger)\big] e^{iz^* a^\dagger} e^{iza}\Big\}.\end{aligned} \tag{4.27}$$

Our aim is to express each of the nine terms on the right-hand side of (4.27) in terms of χ_N and its derivatives with respect to (iz^*) and (iz). For two of the nine terms this can be achieved directly; we may write

$$\mathrm{tr}\left(a\rho a^\dagger e^{iz^* a^\dagger} e^{iza}\right) = \mathrm{tr}\left(\rho a^\dagger e^{iz^* a^\dagger} e^{iza} a\right) = \frac{\partial^2}{\partial(iz^*)\partial(iz)}\chi_N, \tag{4.28}$$

where we simply used the cyclic property of the trace. The remaining seven terms require a little more algebraic manipulation; but the goal is always the same – to rearrange the terms inside the trace so that $a^\dagger$ is to the left of $e^{iz^* a^\dagger}$ and a is to the right of e^{iza}. Then, $a^\dagger$ and a can be brought down from the exponentials by differentiation with respect to (iz^*) and (iz), respectively. Generally, the rearrangement may require us to pass $a^\dagger$ through the exponential e^{iza}, or a through the exponential $e^{iz^* a^\dagger}$. The details of the manipulations are not important. Eventually they bring us to an equation of motion for $\chi_N(z, z^*, t)$ in the form:

$$\frac{\partial \chi_N}{\partial t} = \left[-(\kappa + i\omega_C)z\frac{\partial}{\partial z} - (\kappa - i\omega_C)z^*\frac{\partial}{\partial z^*} - 2\kappa\bar{n}zz^*\right]\chi_N. \tag{4.29}$$

To pass to an equation of motion for $P(\alpha, \alpha^*, t)$, we use the Fourier transform relation (4.16), and exchange the differential operator in the variables z and z^* for one in the variables α and α^*:

$$\begin{aligned}&\int d^2\alpha\, \frac{\partial P(\alpha,\alpha^*,t)}{\partial t}\, e^{iz^*\alpha^*} e^{iz\alpha}\\ &= \int d^2\alpha\, P(\alpha,\alpha^*,t)\bigg[-(\kappa + i\omega_C)z\frac{\partial}{\partial z} - (\kappa - i\omega_C)z^*\frac{\partial}{\partial z^*}\\ &\qquad - 2\kappa\bar{n}zz^*\bigg] e^{iz^*\alpha^*} e^{iz\alpha}\\ &= \int d^2\alpha\, P(\alpha,\alpha^*,t)\bigg[-(\kappa + i\omega_C)(i\alpha)\frac{\partial}{\partial(i\alpha)} - (\kappa - i\omega_C)(i\alpha^*)\frac{\partial}{\partial(i\alpha^*)}\\ &\qquad -2\kappa\bar{n}\frac{\partial^2}{\partial(i\alpha)\partial(i\alpha^*)}\bigg] e^{iz^*\alpha^*} e^{iz\alpha}.\end{aligned} \tag{4.30}$$

The action of the derivatives on the right-hand side of (4.30) can be moved from the product of exponentials, $e^{iz^*\alpha^*} e^{iz\alpha}$, to $P(\alpha, \alpha^*, t)$ by integrating by parts, assuming that $P(\alpha,\alpha^*,t)$ vanishes sufficiently fast at infinity to justify dropping the boundary terms. Then, (4.30) becomes

$$\int d^2\alpha\, e^{iz^*\alpha^*}e^{iz\alpha}\frac{\partial P}{\partial t} = \int d^2\alpha\, e^{iz^*\alpha^*}e^{iz\alpha}\biggl[(\kappa+i\omega_C)\frac{\partial}{\partial\alpha}\alpha + (\kappa-i\omega_C)\frac{\partial}{\partial\alpha^*}\alpha^* + 2\kappa\bar{n}\frac{\partial^2}{\partial\alpha\partial\alpha^*}\biggr]P. \tag{4.31}$$

After inverting the Fourier transform we arrive at the *Fokker-Planck equation for a cavity mode driven by thermal light in the Glauber-Sudarshan P representation*:

$$\frac{\partial P}{\partial t} = \biggl[(\kappa+i\omega_C)\frac{\partial}{\partial\alpha}\alpha + (\kappa-i\omega_C)\frac{\partial}{\partial\alpha^*}\alpha^* + 2\kappa\bar{n}\frac{\partial^2}{\partial\alpha\partial\alpha^*}\biggr]P. \tag{4.32}$$

The Green function solution to (4.32) describes the decay of the cavity mode from an initial coherent state $|\alpha_0\rangle$ [Eq. (4.5)] to the thermal equilibrium state (4.9). It is given by

$$P(\alpha,\alpha^*,t|\alpha_0,\alpha_0^*,0) = \frac{1}{\pi\bar{n}(1-e^{-2\kappa t})}\exp\biggl[-\frac{|\alpha-\alpha_0 e^{-\kappa t}e^{-i\omega_C t}|^2}{\bar{n}(1-e^{-2\kappa t})}\biggr]. \tag{4.33}$$

$P(\alpha,\alpha^*,t|\alpha_0,\alpha_0^*,0)$ is a two-dimensional Gaussian distribution. Thus, for this example, the P distribution has all the properties of a probability distribution. The mean of the Gaussian gives the oscillating and decaying cavity mode amplitude obtained from the expectation value equation (3.1):

$$\langle a(t)\rangle = \alpha_0 e^{-\kappa t}e^{-i\omega_C t}. \tag{4.34}$$

The variance describes the thermal fluctuations added to the coherent amplitude by the oscillator's interaction with the reservoir [compare (3.3)]:

$$\langle (a^\dagger a)(t)\rangle - \langle a^\dagger(t)\rangle\langle a(t)\rangle = \bar{n}(1-e^{-2\kappa t}). \tag{4.35}$$

Similar Fokker-Planck equations are found using the Q and Wigner representations. The only differences are that where $\bar{n}$ appears in the Fokker-Planck equation in the P representation, $\bar{n}+1$ appears in the Fokker-Planck equation in the Q representation and $\bar{n}+\frac{1}{2}$ appears in the Fokker-Planck equation in the Wigner representation. These differences are explained by the different ordering conventions upon which the different representations are based. In the Q representation the variance of the distribution gives $\langle (aa^\dagger)(t)\rangle$, which in the steady state is $\bar{n}+1$, while in the Wigner representation the variance gives $[\langle (a^\dagger a)(t)\rangle + \langle (aa^\dagger)(t)\rangle]/2$ which in the steady state is $\bar{n}+\frac{1}{2}$.

From the conditional distribution (4.33) multitime averages of the classical phase-space variables can be calculated; these also give information about operator averages. In the P representation they give normal-ordered, time-ordered operator averages – for example ($\tau \geq 0$),

$$\begin{aligned}&\langle a^{\dagger p}(t)\hat{N}(t+\tau)a^{q}(t)\rangle\\&\quad=\int d^2\alpha\int d^2\alpha_0\,\alpha_0^{*p}\alpha_0^{q}N(\alpha,\alpha^*)P(\alpha,\alpha^*,t+\tau;\alpha_0,\alpha_0^*,t),\end{aligned}\tag{4.36a}$$

where

$$P(\alpha,\alpha^*,t+\tau;\alpha_0,\alpha_0^*,t)=P(\alpha,\alpha^*,\tau|\alpha_0,\alpha_0^*,0)P(\alpha_0,\alpha_0^*,t),\tag{4.36b}$$

and $\hat{N}$ is any operator written as a series in normal order. In the Q representation the antinormal-ordered, reverse-time-ordered operator averages are obtained – for example ($\tau \geq 0$),

$$\begin{aligned}&\langle a^{q}(t)\hat{A}(t+\tau)a^{\dagger p}(t)\rangle\\&\quad=\int d^2\alpha\int d^2\alpha_0\,\alpha_0^{*p}\alpha_0^{q}A(\alpha,\alpha^*)Q(\alpha,\alpha^*,t+\tau;\alpha_0,\alpha_0^*,t),\end{aligned}\tag{4.37a}$$

where

$$Q(\alpha,\alpha^*,t+\tau;\alpha_0,\alpha_0^*,t)=Q(\alpha,\alpha^*,\tau|\alpha_0,\alpha_0^*,0)Q(\alpha_0,\alpha_0^*,t),\tag{4.37b}$$

and $\hat{A}$ is any operator written as a series in antinormal order. In the Wigner representation the multitime phase-space averages correspond to quantum averages with a symmetric operator and time ordering.

4.3 Stochastic differential equations

The Fokker-Planck equation has a long history, going back to its use by Fokker in 1915 [4.13], and Planck in 1917 [4.14], to describe Brownian motion. In its traditional context it is an equation for a conditional probability density $P(\boldsymbol{x},t|\boldsymbol{x}_0,0)$ of the form

$$\begin{aligned}&\frac{\partial P(\boldsymbol{x},t|\boldsymbol{x}_0,0)}{\partial t}\\&\quad=\left(-\sum_{i=1}^{n}\frac{\partial}{\partial x_i}A_i(\boldsymbol{x})+\frac{1}{2}\sum_{i,j=1}^{n}\frac{\partial^2}{\partial x_i\partial x_j}D_{ij}(\boldsymbol{x})\right)P(\boldsymbol{x},t|\boldsymbol{x}_0,0),\end{aligned}\tag{4.38}$$

where $\boldsymbol{x}$ is a vector of n random variables $x_1,\ldots,x_n$, and the $A_i(\boldsymbol{x})$ and $D_{ij}(\boldsymbol{x})$ are general functions of these variables; the matrix $D_{ij}(\boldsymbol{x})$ is symmetric and positive definite by definition. There are many examples in quantum optics where the quantum-classical correspondence leads to an equation with all of the complexity of (4.38) – many dimensions, and difficult nonlinearities in the functions $A_i(\boldsymbol{x})$ and $D_{ij}(\boldsymbol{x})$. Generally it is not possible to solve such a complicated partial differential equation. But in the age of computers, one way to proceed is to use an equivalent set of stochastic differential equations that can be simulated in a Monte Carlo fashion.

We do not have time to say very much about stochastic differential equations. Perhaps the best thing is to just state the set of equations that is equivalent to (4.38) and describe how these are interpreted in an operational manner. A good reference for further reading is the book by Gardiner [4.15]. The *Ito stochastic differential equations equivalent to the multidimensional Fokker-Planck equation* (4.38) are given by

$$d\boldsymbol{x} = \boldsymbol{A}(\boldsymbol{x})dt + \boldsymbol{B}(\boldsymbol{x})d\boldsymbol{W}, \tag{3.39}$$

where the matrix $\boldsymbol{B}(\boldsymbol{x})$ is defined by the decomposition

$$\boldsymbol{D}(\boldsymbol{x}) = \boldsymbol{B}(\boldsymbol{x})\boldsymbol{B}(\boldsymbol{x})^T \tag{4.40}$$

of the positive definite matrix $\boldsymbol{D}(\boldsymbol{x})$; the stochasticity, or randomness, enters through $d\boldsymbol{W}$, which is a vector of Wiener increments. In practice we can interpret (4.39) as an Euler algorithm for integrating a set of differential equations; at each time step, $d\boldsymbol{W}$ is a vector of independent Gaussian distributed random numbers with mean zero and variance dt, where dt is the integration time step.

4.4 Linearization and the system size expansion

Little progress would be made with Fokker-Planck equation methods if we relied solely on the good fortune of obtaining equations that can be exactly solved, or on time consuming numerical simulation. In fact, often the quantum-classical correspondence does not lead to a Fokker-Planck equation at all, but to an equation involving partial derivatives to all orders. In such situations progress can only be made using approximations. Most work in quantum optics where the quantum-classical correspondence is used to treat an operator master equation also makes use of a system size expansion. This approximation removes derivatives beyond the second order and, in general, also removes nonlinearities – the $A_i(\boldsymbol{x})$ become linear functions of $\boldsymbol{x}$ and the $D_{ij}(\boldsymbol{x})$ become constants. The solution to the resulting linear Fokker-Planck equation is a multidimensional Gaussian distribution from which any desired statistical quantity is fairly readily derived. We look at the system size expansion for a one-dimensional system.

Our discussion is based on the systematic treatment of fluctuations in classical stochastic systems worked out by Van Kampen [4.16]. We begin with the *generalized Fokker-Planck equation*, or what is known in classical stochastic theory as the *Kramers-Moyal expansion* [4.17, 4.18]:

$$\frac{\partial P}{\partial t} = \sum_{k=1}^{\infty} \frac{(-1)^k}{k!} \left(\frac{\partial}{\partial x}\right)^k (a_k(x)P). \tag{4.41}$$

This equation is formally equivalent to the master equation for a classical jump process. It also provides a general form (in one dimension) for the equation of motion for the phase-space distribution obtained via the quantum-classical correspondence. Two difficulties with this equation generally have to be addressed: First, the appearance of derivatives beyond second order. Second, even if these higher-order derivatives are dropped, this will generally leave nonlinearities, which for a multidimensional problem almost certainly make the Fokker-Plank equation impossible to solve. Both of these difficulties can often be removed on the basis of a "small noise" approximation.

The central idea is that the distribution drifts along some trajectory in phase space determined by its time-dependent mean, while simultaneously evolving a "small" width describing fluctuations about the mean. For sufficiently small noise it seems reasonable that this distribution be approximated by a narrow Gaussian; we see in (4.33) that Gaussian distributions are obtained from linear Fokker-Planck equations. The system size expansion follows a systematic path from (4.42) to such a description, basing its development on an expansion in terms of a small parameter related to the inverse of the system "size". The systematic approach offered by the system size expansion leads in a single step to a *linear* Fokker-Planck equation, simultaneously taking care of both of the difficulties mentioned above. This is the consistent thing to do, rather than simply truncating derivatives beyond second order and accepting the nonlinear Fokker-Planck equation that results. As will become clear below, retaining the nonlinearity after truncation brings corrections to the linearized form of the Fokker-Planck which are of the same order as terms which have already been dropped. It is therefore inconsistent not to linearize as well as truncate.

We must look for an expansion parameter which can take us to the limit of zero fluctuations. What is the rationale for expecting such a limiting procedure to be possible? How can the limit be taken formally? Our interest is with intrinsic fluctuations arising in the microscopic quantum processes that govern the interaction of light with matter. The quantized, or discrete, nature of this interaction is the fundamental source of the fluctuations: photon numbers change discretely, and material states follow suit as photons are exchanged with the optical field. If the number of quanta in the field and the number of interacting material states are large, we might expect the fluctuations associated with individual transitions to be small on the scale of the average behavior. Let us imagine we can scale the "size" of a given system with some parameter Ω, to obtain a family of systems, all with the same average behavior, but whose fluctuations decrease relative to the mean as Ω is increased. Let x specify a state in microscopic units (numbers of photons, for example), which therefore scales with system size, and let $\bar{x}$

specify the macroscopic state whose average does not change with Ω. We propose a scaling relationship

$$x = \Omega^p \bar{x}. \tag{4.42}$$

This is a generalization of the relationship postulated for a classical jump process [4.15]. In that relationship $p = 1$. We need the generalization specifically to include the case $p = 1/2$ which is appropriate for optical field amplitudes.

Consider the example of an optical field amplitude. Let x be the amplitude of an optical cavity mode, in units such that x^2 measures the number of photons in the cavity; thus, x corresponds to the variable α in (4.32) – forget for the moment the two-dimensional character of the field. This cavity mode interacts with some intracavity medium. The relevant quantity for describing this interaction at the macroscopic level is not the photon number, but the energy density in the medium. We therefore choose $\bar{x}$ to be scaled so that $\bar{x}^2 \sim 1$ corresponds to energy densities in the range typical of the behavior to be studied (for example, the saturation of a two-state atom, the turn on of a parametric oscillator). The size of the cavity can be scaled up, increasing the photon number x^2 corresponding to any fixed energy density $\bar{x}^2$. If n_0 is the photon number at each cavity size corresponding to the reference energy density $\bar{x}^2 = 1$, we would write (4.42) as

$$x = n_0^{1/2} \bar{x}.$$

$\Omega \equiv n_0$ is a reference photon number and $p = 1/2$.

For a second example let x correspond to the inversion of a two-state medium. The relevant quantity for describing the macroscopic properties of the medium is the inversion density, giving the number of atoms per unit volume available for absorption or emission. Define $\bar{x}$ as the inversion density divided by the atomic density N/V (for N atoms uniformly distributed in a volume V). Systems of increasing size, with fixed atomic density and inversion density $\bar{x}$, have

$$x = N\bar{x}.$$

In this case $\Omega \equiv N$ is a number of atoms and $p = 1$.

The system size expansion now works as follows. We assume that as Ω increases, some mean motion $\bar{x}_0(t)$ is preserved, while fluctuations about this mean decrease. We write

$$\bar{x} = \bar{x}_0(t) + \Omega^{-q}\xi, \tag{4.43a}$$

and introduce the change of variable

$$x = \Omega^p \bar{x}_0(t) + \Omega^{p-q}\xi. \tag{4.43b}$$

The new variable ξ is to be of the same order as $\bar{x}_0(t)$, and q must be determined self-consistently to ensure that this is so from the description of the fluctuations provided by the generalized Fokker-Planck equation (4.42). Setting

$$\bar{P}(\xi,t) = \Omega^{p-q} P\big(\Omega^p \bar{x}_0(t) + \Omega^{p-q}\xi, t\big), \tag{4.44}$$

the generalized Fokker-Planck equation becomes

$$\begin{aligned}\frac{\partial \bar{P}}{\partial t} &= \Omega^{p-q}\left(\frac{\partial P}{\partial \bar{x}_0(t)}\frac{d\bar{x}_0(t)}{dt} + \frac{\partial P}{\partial t}\right)\\ &= \Omega^q \frac{\partial \bar{P}}{\partial \xi}\frac{d\bar{x}_0(t)}{dt} + \sum_{k=1}^{\infty}\frac{(-1)^k}{k!}\left(\Omega^{q-p}\frac{\partial}{\partial \xi}\right)^k (a_k \bar{P}).\end{aligned}$$

Assuming $P(x,t)$ is normalized with respect to the variable x, $\bar{P}(\xi,t)$ has been defined so that it is normalized with respect to the variable ξ. We now make a Taylor expansion of the functions $a_k(x)$ about the mean motion $\Omega^p \bar{x}_0(t)$:

$$\begin{aligned}\frac{\partial \bar{P}}{\partial t} &= \Omega^q \frac{\partial \bar{P}}{\partial \xi}\frac{d\bar{x}_0(t)}{dt}\\ &\quad - \Omega^{q-p}\frac{\partial}{\partial \xi}\Big[a_1\big(\Omega^p\bar{x}_0(t)\big) + \Omega^{p-q}\xi a_1'\big(\Omega^p\bar{x}_0(t)\big) + \tfrac{1}{2}\Omega^{2(p-q)}\\ &\quad \times \xi^2 a_1''\big(\Omega^p\bar{x}_0(t)\big) + \cdots\Big]\bar{P}\\ &\quad + \frac{1}{2}\Omega^{2(q-p)}\frac{\partial^2}{\partial \xi^2}\Big[a_2\big(\Omega^p\bar{x}_0(t)\big) + \Omega^{p-q}\xi a_2'\big(\Omega^p\bar{x}_0(t)\big) + \tfrac{1}{2}\Omega^{2(p-q)}\\ &\quad \times \xi^2 a_2''\big(\Omega^p\bar{x}_0(t)\big) + \cdots\Big]\bar{P}\\ &\quad + \\ &\quad \vdots\end{aligned} \tag{4.45}$$

where $'$ denotes differentiation with respect to x.

To take things further we need to know how the functions $a_k\big(\Omega^p\bar{x}_0(t)\big)$ scale with Ω. In the context of classical jump processes this scaling can be argued from the dependence of the a_k – the jump moments – on the transition probability for a jump of given length from an initial state x. Our derivation of the Fokker-Planck equation from an operator master equation cannot rely on the same argument; in fact, the scaling adopted for a jump process must be generalized to include variables corresponding to field amplitudes, for which $p = 1/2$ rather than $p = 1$. To cover both values of p we use

$$a_k\big(\Omega^p\bar{x}_0(t)\big) = \Omega^{k(p-1)+1}\bar{a}_k\big(\bar{x}_0(t)\big). \tag{4.46}$$

Then the expansion (4.46) becomes

$$\begin{aligned}\frac{\partial \bar{P}}{\partial t} &= \Omega^q \left[\frac{d\bar{x}_0(t)}{dt} - \bar{a}_1(\bar{x}_0(t))\right]\frac{\partial \bar{P}}{\partial \xi} \\ &\quad - \frac{\partial}{\partial \xi}\xi\left[\bar{a}_1'(\bar{x}_0(t)) + \tfrac{1}{2}\Omega^{-q}\xi\bar{a}_1''(\bar{x}_0(t)) + O(\Omega^{-2q})\right]\bar{P} \\ &\quad + \frac{1}{2}\Omega^{2q-1}\frac{\partial^2}{\partial \xi^2}\left[\bar{a}_2(\bar{x}_0(t)) + \Omega^{-q}\xi\bar{a}_2'(\bar{x}_0(t)) + O(\Omega^{-2q})\right]\bar{P} \\ &\quad + O(\Omega^{3q-2}), \end{aligned} \tag{4.47}$$

where $'$ now denotes differentiation with respect to $\bar{x}$.

We have now reached the point where we impose self-consistency on our expansion; we require that (4.47) produce fluctuations ξ of the order $\bar{x}_0(t)$ in the limit of large Ω, as was assumed in the ansatz (4.43a). To avoid the divergence of the first term on the right-hand side the factor in the square bracket must vanish identically, which requires that

$$\frac{d\bar{x}_0(t)}{dt} = \bar{a}_1(\bar{x}_0(t)). \tag{4.48}$$

This is the *macroscopic law* governing the mean motion of the system. The self-consistency requirement also sets the size of q. Assuming that $\bar{a}_1'(\bar{x}_0(t))$ and $\bar{a}_2(\bar{x}_0(t))$ are both nonzero, we must clearly choose $q = 1/2$. Then the right-hand side of (4.47) becomes an expansion in powers of $\Omega^{-1/2}$, and in the limit of large Ω the dominant terms give the linear Fokker-Planck equation

$$\frac{\partial \bar{P}}{\partial t} = \left[-\bar{a}_1'(\bar{x}_0(t))\frac{\partial}{\partial \xi}\xi + \frac{1}{2}\bar{a}_2(\bar{x}_0(t))\frac{\partial^2}{\partial \xi^2}\right]\bar{P}. \tag{4.49}$$

Given a trajectory $\bar{x}_0(t)$ satisfying (4.48), equation (4.49) can be solved for a Gaussian distribution which drifts along this trajectory, accumulating a width as it goes, given by integration over the time-dependent diffusion. For (4.49) the Gaussian solution is

$$\bar{P}(\xi, t) = \frac{1}{\sqrt{2\pi}\sigma(t)} \exp\left[-\frac{\left(\xi - \langle\xi(t)\rangle\right)^2}{2\sigma^2(t)}\right], \tag{4.50}$$

with mean

$$\langle\xi(t)\rangle = \langle\xi(0)\rangle \exp\left[\int_0^t du\, \bar{a}_1'(\bar{x}_0(u))\right], \tag{4.51a}$$

and variance

$$\begin{aligned}\sigma^2(t) &= \exp\left[2\int_0^t du\, \bar{a}_1'(\bar{x}_0(u))\right] \\ &\quad \times \left\{\sigma^2(0) + \int_0^t du\, \exp\left[-2\int_0^u dv\, \bar{a}_1'(\bar{x}_0(v))\right]\bar{a}_2(\bar{x}_0(u))\right\}. \end{aligned} \tag{4.51b}$$

Since the original construction puts the mean motion in $\bar{x}_0(t)$, this solution is to be taken with $\langle\xi(0)\rangle = 0$.

4.5 The degenerate parametric oscillator

We now illustrate the use of the system size expansion for the example of the degenerate parametric oscillator. The master equation is given in (2.63). The phase-space equations of motion corresponding to this master equation in the P, Q, and Wigner representations are summarized by the single equation

$$\frac{\partial F_\sigma}{\partial t} = L_\sigma\left(\alpha, \alpha^*, \beta, \beta^*, \frac{\partial}{\partial\alpha}, \frac{\partial}{\partial\alpha^*}, \frac{\partial}{\partial\beta}, \frac{\partial}{\partial\beta^*}, t\right) F_\sigma, \tag{4.52}$$

where

$$\begin{aligned}
&L_\sigma\left(\alpha, \alpha^*, \beta, \beta^*, \frac{\partial}{\partial\alpha}, \frac{\partial}{\partial\alpha^*}, \frac{\partial}{\partial\beta}, \frac{\partial}{\partial\beta^*}, t\right) \\
&= \frac{\partial}{\partial\alpha}[(\kappa + i\omega_C)\alpha - g\alpha^*\beta] + \frac{\partial}{\partial\alpha^*}[(\kappa - i\omega_C)\alpha^* - g\alpha\beta^*] \\
&\quad + \frac{\partial}{\partial\beta}\left[(\kappa_p + i2\omega_C)\beta + (g/2)\alpha^2 + i\bar{\mathcal{E}}_i e^{-i2\omega_C t}\right] \\
&\quad + \frac{\partial}{\partial\beta^*}\left[(\kappa_p - i2\omega_C)\beta^* + (g/2)\alpha^{*2} - i\bar{\mathcal{E}}_i^* e^{i2\omega_C t}\right] \\
&\quad + \kappa(1-\sigma)\frac{\partial^2}{\partial\alpha\partial\alpha^*} + \kappa_p(1-\sigma)\frac{\partial^2}{\partial\beta\partial\beta^*} \\
&\quad + \sigma(g/2)\left(\frac{\partial^2}{\partial\alpha^2}\beta + \frac{\partial^2}{\partial\alpha^{*2}}\tilde{\beta}^*\right) + \frac{1}{4}(1-|\sigma|)(g/2)\left(\frac{\partial^3}{\partial\alpha^2\partial\beta^*} + \frac{\partial^3}{\partial\alpha^{*2}\partial\beta}\right);
\end{aligned} \tag{4.53}$$

σ takes the values $+1$, 0, and -1, with the definitions

$$\left.\begin{aligned} F_{+1} &\equiv P \\ F_0 &\equiv W \\ F_{-1} &\equiv Q \end{aligned}\right\}. \tag{4.54}$$

Note that in the Wigner representation (4.53) includes derivatives up to third order, while in the P and Q representations only first- and second-order derivatives appear. Nonlinearities appear due to the nonlinear character of the $\chi^{(2)}$ interaction, and it seems unlikely that an exact solution to (4.53) can be found in any of the representations.

To implement the system size expansion we need a scaling for the subharmonic and pump fields in the form (4.43). A classical treatment of the degenerate parametric oscillator tells us that the undepleted pump photon number at threshold is $n_p^{thr} = (\kappa/g)^2$. This is a natural choice for the system size parameter. The powers (p and q) of n_p^{thr} that appear in (4.43b) are to be chosen for self-consistency in the manner just outlined. We do not have time to go through the details of this calculation here; we just state the scaling that works; this has $p = q = 1/2$. With a little fine-tuning to give a simple form to the final equations, for the subharmonic mode we write

$$\sqrt{\xi/2}e^{-i\psi}\alpha = \left(n_p^{thr}\right)^{1/2}\bar{\alpha}, \quad \sqrt{\xi/2}e^{i\psi}\alpha^* = \left(n_p^{thr}\right)^{1/2}\bar{\alpha}^*, \tag{4.55}$$

with

$$\bar{\alpha} = \langle\bar{a}(t)\rangle + \left(n_p^{thr}\right)^{-1/2}z, \tag{4.56a}$$

$$\bar{\alpha}^* = \langle\bar{a}^\dagger(t)\rangle + \left(n_p^{thr}\right)^{-1/2}z^*, \tag{4.56b}$$

where

$$\sqrt{\xi/2}e^{-i\psi}a = \left(n_p^{thr}\right)^{1/2}\bar{a}, \quad \sqrt{\xi/2}e^{i\psi}a^\dagger = \left(n_p^{thr}\right)^{1/2}\bar{a}^\dagger; \tag{4.57}$$

for the pump mode we write

$$e^{-i\psi}\beta = \left(n_p^{thr}\right)^{1/2}\bar{\beta}, \qquad e^{i\psi}\beta^* = \left(n_p^{thr}\right)^{1/2}\bar{\beta}^*, \tag{4.58}$$

with

$$\bar{\beta} = \langle\bar{b}(t)\rangle + \left(n_p^{thr}\right)^{-1/2}w, \tag{4.59a}$$

$$\bar{\beta}^* = \langle\bar{b}^\dagger(t)\rangle + \left(n_p^{thr}\right)^{-1/2}w^*, \tag{4.59b}$$

where

$$e^{-i\psi}b = \left(n_p^{thr}\right)^{1/2}\bar{b}, \qquad e^{i\psi}b^\dagger = \left(n_p^{thr}\right)^{1/2}\bar{b}^\dagger. \tag{4.60}$$

Here $\xi = \kappa/\kappa_p$ and ψ is a phase that depends on such things as the phase of the pump field.

In terms of the scaled variables the phase-space distributions are defined by

$$\bar{F}_\sigma(z, z^*, w, w^*, t) = \xi^{-1}F_\sigma\big(\alpha(z,t), \alpha^*(z^*,t), \beta(w,t), \beta^*(w^*,t), t\big), \tag{4.61}$$

and satisfy the equation of motion

$$\begin{aligned}
\frac{\partial\bar{F}_\sigma}{\partial t} &= \xi^{-1}\left(\frac{\partial F_\sigma}{\partial\alpha}\frac{\partial\alpha}{\partial t} + \frac{\partial F_\sigma}{\partial\alpha^*}\frac{\partial\alpha^*}{\partial t} + \frac{\partial F_\sigma}{\partial\beta}\frac{\partial\beta}{\partial t} + \frac{\partial F_\sigma}{\partial\beta^*}\frac{\partial\beta^*}{\partial t} + \frac{\partial F_\sigma}{\partial t}\right) \\
&= \left(n_p^{thr}\right)^{1/2}\left(\frac{\partial\bar{F}_\sigma}{\partial z}\frac{d\langle\bar{a}(t)\rangle}{dt} + \text{c.c.}\right) + \left(n_p^{thr}\right)^{1/2}\left(\frac{\partial\bar{F}_\sigma}{\partial w}\frac{d\langle\bar{b}(t)\rangle}{dt} + \text{c.c.}\right) \\
&\quad + \frac{\partial}{\partial t}\left(\xi^{-1}F_\sigma\right) \\
&= \left(n_p^{thr}\right)^{1/2}\left(\frac{\partial\bar{F}_\sigma}{\partial z}\frac{d\langle\bar{a}(t)\rangle}{dt} + \text{c.c.}\right) + \left(n_p^{thr}\right)^{1/2}\left(\frac{\partial\bar{F}_\sigma}{\partial w}\frac{d\langle\bar{b}(t)\rangle}{dt} + \text{c.c.}\right) \\
&\quad + \bar{L}_\sigma\left(z, z^*, w, w^*, \frac{\partial}{\partial z}, \frac{\partial}{\partial z^*}, \frac{\partial}{\partial w}, \frac{\partial}{\partial w^*}, t\right)\bar{F}_\sigma,
\end{aligned} \tag{4.62}$$

where

$$
\begin{aligned}
&\bar{L}_\sigma\left(z, z^*, w, w^*, \frac{\partial}{\partial z}, \frac{\partial}{\partial z^*}, \frac{\partial}{\partial w}, \frac{\partial}{\partial w^*}, t\right) \\
&= L_\sigma\left(\alpha(z,t), \alpha^*(z^*,t), \beta(w,t), \beta^*(w^*,t), \sqrt{\xi/2}e^{-i\psi}\frac{\partial}{\partial z},\right. \\
&\qquad \left.\sqrt{\xi/2}e^{i\psi}\frac{\partial}{\partial z^*}, e^{-i\psi}\frac{\partial}{\partial w}, e^{i\psi}\frac{\partial}{\partial w^*}, t\right).
\end{aligned} \tag{4.63}
$$

The macroscopic law governing the mean behavior and the Fokker-Planck equation that describes fluctuations about the mean are identified after we substitute the explicit form for L_σ from (4.53). We obtain

$$
\begin{aligned}
&\frac{\partial \bar{F}_\sigma}{\partial t} \\
&= \left(n_p^{thr}\right)^{1/2}\frac{\partial \bar{F}_\sigma}{\partial z}\left[\frac{d\langle\bar{a}(t)\rangle}{dt} + (\kappa + i\omega_C)\langle\bar{a}(t)\rangle - \kappa\langle\bar{a}^\dagger(t)\rangle\langle\bar{b}(t)\rangle\right] + \text{c.c.} \\
&\quad + \left(n_p^{thr}\right)^{1/2}\frac{\partial \bar{F}_\sigma}{\partial w}\left[\frac{d\langle\bar{b}(t)\rangle}{dt} + (\kappa_p + i2\omega_C)\langle\bar{b}(t)\rangle\right. \\
&\quad \left. + \kappa_p\left(\langle\bar{a}(t)\rangle^2 - \lambda e^{-i2\omega_C t}\right)\right] + \text{c.c.} \\
&\quad + \left\{\frac{\partial}{\partial z}\left[(\kappa + i\omega_C)z - \kappa\left(\langle\bar{a}^\dagger(t)\rangle w + \langle\bar{b}(t)\rangle z^* + \left(n_p^{thr}\right)^{-1/2} z^* w\right)\right] + \text{c.c.}\right. \\
&\quad + \frac{\partial}{\partial w}\left[(\kappa_p + i2\omega_C)w + \kappa_p\left(2\langle\bar{a}(t)\rangle z + \left(n_p^{thr}\right)^{-1/2} z^2\right)\right] + \text{c.c.} \\
&\quad + \frac{1}{4}\xi\sigma\kappa\left[\frac{\partial^2}{\partial z^2}\left(\langle\bar{b}(t)\rangle + \left(n_p^{thr}\right)^{-1/2} w\right) + \text{c.c.}\right] \\
&\quad + (\xi/2)\kappa(1-\sigma)\frac{\partial^2}{\partial z \partial z^*} + \kappa_p(1-\sigma)\frac{\partial^2}{\partial w \partial w^*} \\
&\quad \left. + \left(n_p^{thr}\right)^{-1/2}\frac{1}{16}\xi(1-|\sigma|)\kappa\left(\frac{\partial^3}{\partial z^2 \partial w^*} + \text{c.c.}\right)\right\}\bar{F}_\sigma,
\end{aligned} \tag{4.64}
$$

where $\lambda = (g/\kappa\kappa_p)|\bar{\mathcal{E}}_i|$. To prevent a divergence for $n_p^{thr} \to \infty$ the terms multiplying $\left(n_p^{thr}\right)^{1/2}$ must vanish. This gives the *degenerate parametric oscillator equations without fluctuations:*

$$
\kappa^{-1}\frac{d\langle\tilde{\bar{a}}\rangle}{dt} = -\langle\tilde{\bar{a}}\rangle + \langle\tilde{\bar{a}}^\dagger\rangle\langle\tilde{\bar{b}}\rangle, \tag{4.65a}
$$

$$
\kappa^{-1}\frac{d\langle\tilde{\bar{a}}^\dagger\rangle}{dt} = -\langle\tilde{\bar{a}}^\dagger\rangle + \langle\tilde{\bar{a}}\rangle\langle\tilde{\bar{b}}^\dagger\rangle, \tag{4.65b}
$$

$$
\kappa_p^{-1}\frac{d\langle\tilde{\bar{b}}\rangle}{dt} = -\langle\tilde{\bar{b}}\rangle - \langle\tilde{\bar{a}}\rangle^2 + \lambda, \tag{4.65c}
$$

$$
\kappa_p^{-1}\frac{d\langle\tilde{\bar{b}}^\dagger\rangle}{dt} = -\langle\tilde{\bar{b}}^\dagger\rangle - \langle\tilde{\bar{a}}^\dagger\rangle^2 + \lambda, \tag{4.65d}
$$

where we have removed the free oscillation of the field amplitudes with the transformation

$$\langle\tilde{a}\rangle = e^{i\omega_C t}\bar{a}, \qquad \langle\tilde{\bar{a}}^\dagger\rangle = e^{-i\omega_C t}\bar{a}^\dagger, \tag{4.66a}$$

$$\tilde{\bar{b}} = e^{i2\omega_C t}\bar{b}, \qquad \tilde{\bar{b}}^\dagger = e^{-i2\omega_C t}\bar{b}^\dagger. \tag{4.66b}$$

From the remaining terms in (4.64), dropping terms of order $\left(n_p^{thr}\right)^{-1/2}$, the *linearized Fokker-Planck equation for the degenerate parametric oscillator* reads

$$\begin{aligned}\frac{\partial\tilde{\bar{F}}_\sigma}{\partial t}\\ =\Bigg\{&\kappa\frac{\partial}{\partial\tilde{z}}\left(\tilde{z}-\langle\tilde{\bar{a}}^\dagger(t)\rangle\tilde{w}-\langle\tilde{\bar{b}}(t)\rangle\tilde{z}^*\right)+\kappa\frac{\partial}{\partial\tilde{z}^*}\left(\tilde{z}^*-\langle\tilde{\bar{a}}(t)\rangle\tilde{w}^*-\langle\tilde{\bar{b}}^\dagger(t)\rangle\tilde{z}\right)\\ &+\kappa_p\frac{\partial}{\partial\tilde{w}}\left(\tilde{w}+2\langle\tilde{\bar{a}}(t)\rangle\tilde{z}\right)+\kappa_p\frac{\partial}{\partial\tilde{w}^*}\left(\tilde{w}^*+2\langle\tilde{\bar{a}}^\dagger(t)\rangle\tilde{z}^*\right)\\ &+\frac{1}{4}\xi\sigma\kappa\left(\frac{\partial^2}{\partial\tilde{z}^2}\langle\tilde{\bar{b}}(t)\rangle+\frac{\partial^2}{\partial\tilde{z}^{*2}}\langle\tilde{\bar{b}}^\dagger(t)\rangle\right)\\ &+(\xi/2)\kappa(1-\sigma)\frac{\partial^2}{\partial\tilde{z}\partial\tilde{z}^*}+\kappa_p(1-\sigma)\frac{\partial^2}{\partial\tilde{w}\partial\tilde{w}^*}\Bigg\}\tilde{\bar{F}}_\sigma,\end{aligned} \tag{4.67}$$

where

$$\tilde{\bar{F}}_\sigma(\tilde{z},\tilde{z}^*,\tilde{w},\tilde{w}^*,t)=\bar{F}_\sigma\big(z(\tilde{z},t),z^*(\tilde{z}^*,t),w(\tilde{w},t),w^*(\tilde{w}^*,t),t\big), \tag{4.68}$$

with

$$z = e^{-i\omega_C t}\tilde{z}, \qquad z^* = e^{i\omega_C t}\tilde{z}^*, \tag{4.69a}$$

$$w = e^{-i2\omega_C t}\tilde{w}, \qquad w^* = e^{i2\omega_C t}\tilde{w}^*. \tag{4.69b}$$

If we are interested, for example, in quantum fluctuations below threshold, we set $\langle\tilde{\bar{a}}(t)\rangle = \langle\tilde{\bar{a}}^\dagger(t)\rangle = 0$, and $\langle\tilde{\bar{b}}(t)\rangle = \langle\tilde{\bar{b}}^\dagger(t)\rangle = \lambda$ in (4.67). The linearized Fokker-Planck equation is then separable, with a solution in the form

$$\tilde{\bar{F}}_\sigma(\tilde{z},\tilde{z}^*,\tilde{w},\tilde{w}^*,t)=\tilde{\bar{X}}_\sigma(\tilde{z}_1,t)\tilde{\bar{Y}}_\sigma(\tilde{z}_2,t)\tilde{\bar{U}}_\sigma(\tilde{w}_1,t)\tilde{\bar{V}}_\sigma(\tilde{w}_2,t), \tag{4.70}$$

where

$$\tilde{z} = \tilde{z}_1 + i\tilde{z}_2, \qquad \tilde{z}^* = \tilde{z}_1 - i\tilde{z}_2, \tag{4.71a}$$

$$\tilde{w} = \tilde{w}_1 + i\tilde{w}_2, \qquad \tilde{w}^* = \tilde{w}_1 - i\tilde{w}_2. \tag{4.71b}$$

Fluctuations in the subharmonic field are described by the equations

$$\kappa^{-1}\frac{\partial\tilde{\bar{X}}_\sigma}{\partial t}=\left\{(1-\lambda)\frac{\partial}{\partial\tilde{z}_1}\tilde{z}_1+\frac{1}{8}\xi\big[1-\sigma(1-\lambda)\big]\frac{\partial^2}{\partial\tilde{z}_1^2}\right\}\tilde{\bar{X}}_\sigma, \tag{4.72a}$$

$$\kappa^{-1}\frac{\partial\tilde{\bar{Y}}_\sigma}{\partial t}=\left\{(1+\lambda)\frac{\partial}{\partial\tilde{z}_2}\tilde{z}_2+\frac{1}{8}\xi\big[1-\sigma(1+\lambda)\big]\frac{\partial^2}{\partial\tilde{z}_2^2}\right\}\tilde{\bar{Y}}_\sigma, \tag{4.72b}$$

and fluctuations in the pump field are described by the equations

$$\kappa_p^{-1}\frac{\partial \bar{\bar{U}}_\sigma}{\partial t} = \left\{ \frac{\partial}{\partial \tilde{w}_1}\tilde{w}_1 + \frac{1}{4}(1-\sigma)\frac{\partial^2}{\partial \tilde{w}_1^2} \right\} \bar{\bar{U}}_\sigma, \tag{4.73a}$$

$$\kappa_p^{-1}\frac{\partial \bar{\bar{V}}_\sigma}{\partial t} = \left\{ \frac{\partial}{\partial \tilde{w}_2}\tilde{w}_2 + \frac{1}{4}(1-\sigma)\frac{\partial^2}{\partial \tilde{w}_2^2} \right\} \bar{\bar{V}}_\sigma. \tag{4.73b}$$

We will discuss the physics contained in these equations in a future lecture.

References

[4.1] H. Risken, *The Fokker Planck Equation*, Springer: Berlin, 1984.
[4.2] E. P. Wigner, *Phys. Rev.* **40**, 749 (1932).
[4.3] R. J. Glauber, *Phys. Rev.* **131**, 2766 (1963).
[4.4] E. C. G. Sudarshan, *Phys. Rev. Lett.* **10**, 277 (1963).
[4.5] R. J. Glauber, *Phys. Rev. Lett.* **10**, 84 (1963).
[4.6] R. J. Glauber, *Phys. Rev.* **130**, 2529 (1963).
[4.7] M. J. Lighthill, *Fourier Analysis and Generalized Functions*, Cambridge University Press: Cambridge, 1960.
[4.8] L. Schwartz, *Théorie des Distributions*, Hermann: Paris, Vol. I, 1950, Vol. II, 1951 (2nd edition: 1957/1959).
[4.9] H. Bremermann, *Distributions, Complex Variables, and Fourier Transforms*, Addison-Wesley: Reading, Massachusetts, 1965.
[4.10] J. R. Klauder and E. C. G. Sudarshan, *Fundamentals of Quantum Optics*, Benjamin: New York, 1968, pp. 178ff.
[4.11] W. Feller, *An Introduction to Probability Theory and its Applications*, Vol. II, Wiley: New York, 1966 (2nd edition: 1971), Chapt. XV.
[4.12] H. Weyl, *The Theory of Groups and Quantum Mechanics*, Dover: New York, 1950, pp. 272ff.
[4.13] A. D. Fokker, *Ann. Phys.* (Leipzig), **43**, 310 (1915).
[4.14] M. Planck, *Sitzungsber. Preuss. Akad. Wiss. Phys. Math. Kl.*, **325** (1917).
[4.15] C. W. Gardiner, *Handbook of Stochastic Methods for Physics, Chemistry and the Natural Sciences*, Springer: Berlin, 1983.
[4.16] N. G. Van Kampen, *Stochastic Processes in Physics and Chemistry*, North-Holland: Amsterdam, 1981.
[4.17] H. A. Kramers, *Physica*, **7**, 284 (1940).
[4.18] J. E. Moyal, *J. R. Stat. Soc.*, **11**, 151 (1949).

Lecture 5 – Photoelectric Detection I

In the last four lectures we have reviewed a lot of standard material in quantum optics. In brief we have seen how an operator master equation provides a compact description of a photoemissive source; we have seen how to construct the radiated fields in terms of source operators; and we have seen how the master equation can be analyzed so that we can calculate things like correlation functions for the emitted light. The next two lectures are going to form a bridge between this standard material and the novel formulation of master equation dynamics that will occupy us in the final four lectures. The bridge is built on an understanding of the way in which optical fields are observed. Photoemissive sources are eventually observed by photoelectric detectors. We will spend the next two lectures discussing various aspects of photoelectric detection.

5.1 Photoelectron counting for a constant intensity classical field

When we use a photonmultiplier as a detector we see pulses generated by individual photoelectron emissions. A great deal of information about the statistics of the field that is detected can be obtained by simply counting these pulses over some time interval T. The number of pulses counted will vary if such an experiment is repeated over and over again, because the process of photoelectric emission is fundamentally probabilistic. Two factors contribute to the probabilistic character of the photoelectric emissions: statistical fluctuations in the detected field, and the quantum nature of the interaction between the detector and the field, which only permits us to obtain probabilities for photoelectric emission – not predictions that a photoelectron will definitely be emitted at this particular time or at that particular time. The probability density $p(n,t,T)$ for counting n photoelectrons in the interval $(t,t+T]$ is called the *photoelectron counting distribution*. To separate the contributions from field statistics and the emission process, we first calculate the photoelectron counting distribution for a constant intensity classical field, where, by definition, the effects of field statistics are absent.

We consider a beam of light with frequency ω incident on a phototube that intersects a cross-sectional area A of the beam over which the cycle-averaged intensity $\bar{I}$ is uniform. We can express the incident power at the detector in terms of the photon energy $\hbar\omega$ multiplied by the average number of photons per second entering the detector. This gives the relationship

$$\begin{pmatrix}\text{average number of}\\ \text{photons entering the}\\ \text{detector per second}\end{pmatrix} = \frac{A\bar{I}}{\hbar\omega}, \tag{5.1}$$

If the detector counts for a time T, we expect that

$$\begin{pmatrix}\text{average number of}\\ \text{photons counted}\\ \text{in the time } T\end{pmatrix} = \frac{A\bar{I}}{\hbar\omega}T, \tag{5.2}$$

where we have assumed that the detector is able to record a photoelectron for every available photon. Of course, in practice this assumption is not correct. The detector has a *quantum efficiency* η, which is a number between zero and one designating the proportion of available photons that, on average, actually result in a photoelectron. The quantum efficiency will depend in a complicated way on the detector design, and we may regard it as an empirical parameter, not something to be calculated from first principles. Taking the quantum efficiency into account, we write

$$\begin{pmatrix}\text{average number of}\\ \text{photons counted}\\ \text{in the time } T\end{pmatrix} = \eta\frac{A\bar{I}}{\hbar\omega}T = \xi\bar{I}T, \tag{5.3}$$

where

$$\xi = \eta\frac{A}{\hbar\omega}. \tag{5.4}$$

Note that $\xi\bar{I}$ is a photon flux.

Now the fundamental notion given to us by the quantum treatment of the photoelectric effect is a probability for photoelectric emission in some infinitesimal interaction time Δt. Then the number of photoelectrons counted in a finite time interval of duration T is characterized by a probability distribution derived from this emission probability. Let us divide the total time T into $N = T/\Delta t \gg 1$ subintervals. If the incident field is a constant intensity classical field, all the subintervals are equivalent (we assume that the detector recovers infinitely fast after emitting a photoelectron so that it is available, instantly, to emit another). We then calculate the probability for counting n photoelectrons during the interval T from the statistics of N independent coin tosses – "heads" indicates that one photoelectron is emitted in a given subinterval Δt; "tails" indicates that no photoelectron is emitted during that subinterval. The probability p for tossing "heads" is proportional to the photon flux illuminating the detector; the probability q for tossing "tails" is given by $1 - p$; we write

$$p = \xi \bar{I} \Delta t, \tag{5.5a}$$
$$q = 1 - \xi \bar{I} \Delta t. \tag{5.5b}$$

Note that it is always possible to eliminate events in which two or more photoelectrons are emitted in the interval Δt by simply making the interval sufficiently short. Now $p(n,t,T)$ is the probability for tossing n "heads" and $N-n$ "tails:"

$$\begin{aligned} p(n,t,T) &= \frac{N(N-1)\cdots(N-n+1)}{n!}(\xi\bar{I}\Delta t)^n(1-\xi\bar{I}\Delta t)^{N-n} \\ &= \left(1-\frac{1}{N}\right)\cdots\left(1-\frac{n-1}{N}\right)\frac{(\xi\bar{I}N\Delta t)^n}{n!}(1-\xi\bar{I}\Delta t)^{N-n}. \end{aligned}$$

The combinatorial factor accounts for the different orders in which n "heads" and $N-n$ "tails" can appear. We want Δt to be very small. We therefore take the limit $N \to \infty$, $\Delta t \to 0$, with $N\Delta t = T$ constant. In this limit,

$$(1-\xi\bar{I}\Delta t)^{N-n} \to \exp(-\xi\bar{I}T), \tag{5.6}$$

and we obtain

$$p(n,t,T) = \frac{(\xi\bar{I}T)^n}{n!}\exp(-\xi\bar{I}T). \tag{5.7}$$

This is a Poisson distribution. It is independent of the time t at the start of the counting interval for the obvious reason that a constant intensity field should produce the same counting distribution for all intervals of the same duration, regardless of the origin in time.

The Poisson photoelectron counting distribution is peaked about velues of n in the range $\bar{n}-\sqrt{\bar{n}}$ to $\bar{n}+\sqrt{\bar{n}}$, and the deviation from the mean becomes small, relatively speaking – $\sim 1/\sqrt{\bar{n}}$ – as the mean $\bar{n}$ becomes large. The Poisson distribution of photoelectric counts for constant intensity light is the origin of what electrical engineers call *shot noise*. We will have more to say about shot noise when we discuss the detection of squeezed light in the next lecture. We simply note for the moment that the signal-to-noise ratio for shot noise can be improved by increasing $\bar{n}$ – by using higher intensities.

5.2 Photoelectron counting for a general classical field

Optical fields with constant intensity are not very interesting. But we have learned something by considering them first. We have seen that there is a Poisson distribution of photoelectrons emitted over a fixed interval of time, even when there is no uncertainty – no fluctuations – in the field being detected. Thus, we have separated the uncertainty in the count of photoelectrons due to the quantum nature of the emission process from any explicit fluctuations in the field. We now introduce explicit fluctuations.

We can readily generalize (5.7) to a situation in which $\bar{I}(t)$ is essentially constant over each counting interval $(t, t+T]$, but changes (slowly) between the counting intervals that are put together to form a complete ensemble of measurements. This corresponds to situations in which the counting time T is short compared to the intensity correlation time for the detected light, as illustrated in Fig. 5.1. Under these conditions we can write

$$\begin{aligned} p(n,t,T) &= \left\langle \frac{(\xi \bar{I}T)^n}{n!} \exp(-\xi \bar{I}T) \right\rangle \\ &= \int_0^\infty d\bar{I}\, P(\bar{I}) \frac{(\xi \bar{I}T)^n}{n!} \exp(-\xi \bar{I}T), \end{aligned} \tag{5.8}$$

where $P(\bar{I})$ is the probability distribution for the sampled light intensities.

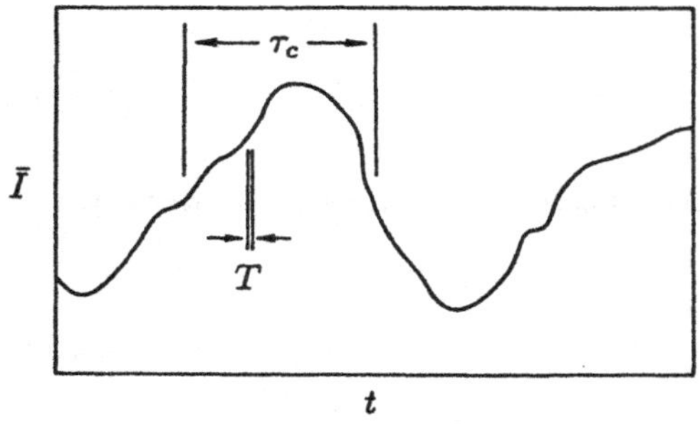

Fig. 5.1. Comparison between the photoelectron counting time T and the intensity correlation time τ_c when (5.8) is valid.

In general the photoelectron counting time T will not be much less then the intensity correlation time. The change that this brings to the photoelectron counting distribution is not unexpected; we keep (5.8), but make the replacement

$$\xi \bar{I}T \to \xi \int_t^{t+T} dt' \bar{I}(t'). \tag{5.9}$$

Thus, for a general fluctuating classical field we have

$$p(n,t,T) = \left\langle \frac{\left(\xi \int_t^{t+T} dt' \bar{I}(t')\right)^n}{n!} \exp\left(-\xi \int_t^{t+T} dt' \bar{I}(t')\right) \right\rangle. \tag{5.10}$$

This is sometimes referred to as *the Mandel photoelectron counting formula*, in recognition of Mandel's early derivation of the result [5.1]. A derivation of (5.9) is given in Loudon's text [5.2]. The way to understand (5.10) is to visualize an intensity fluctuating in time as in Fig. 5.1. For each sampling interval $(t, t+T]$ the integral in (5.9) calculates the accumulated number of photons (multiplied by η) incident on the detector – the integral of a time-varying photon flux. If the intensity fluctuates in a stochastic way this integral is a random variable; thus, we need the ensemble average taken in

(5.10). Note that an ensemble average is taken here to get the photoelectron counting distribution. A second average, taken against this distribution, is needed to calculate such things as the mean and variance of the photoelectron number.

5.3 Moments of the counting distribution

To simplify the notation we write the photoelectron counting distribution in the form

$$p(n,t,T) = \left\langle \frac{\Omega^n}{n!} e^{-\Omega} \right\rangle, \tag{5.11a}$$

where

$$\Omega(t,T) \equiv \xi \int_t^{t+T} dt' \bar{I}(t'). \tag{5.11b}$$

Moments of this distribution are easy to calculate using the *moment generating function*

$$\phi(y) \equiv \left\langle \exp\left(y \int_t^{t+T} dt' \bar{I}(t') \right) \right\rangle. \tag{5.12}$$

This moment generating function is particularly suited for calculating the factorial moments

$$n^{(r)}(t,T) = \sum_{n=0}^{\infty} n(n-1)\cdots(n-r+1) p(n,t,T). \tag{5.13}$$

These are obtained by taking derivatives, a procedure that is similar to the one used in (4.14) to obtain operator averages from the characteristic function:

$$\begin{aligned}
n^{(r)} &= \sum_{n=0}^{\infty} n(n-1)\cdots(n-r+1) \left\langle \frac{\Omega^n}{n!} e^{-\Omega} \right\rangle \\
&= (-1)^r \sum_{n=r}^{\infty} \left\langle \frac{\Omega^{n-r}}{(n-r)!} \frac{d^r}{dx^r} e^{-x\Omega} \right\rangle \Bigg|_{x=1} \\
&= (-1)^r \frac{d^r}{dx^r} \sum_{k=0}^{\infty} \left\langle \frac{\Omega^k}{k!} e^{-x\Omega} \right\rangle \Bigg|_{x=1} \\
&= (-1)^r \frac{d^r}{dx^r} \left\langle e^{(1-x)\Omega} \right\rangle \Bigg|_{x=1}
\end{aligned}$$

Setting $y = 1 - x$ we have

$$n^{(r)} = \frac{d^r}{dy^r}\langle e^{y\Omega}\rangle\bigg|_{y=0} = \langle \Omega^r \rangle. \tag{5.14}$$

The factorial moments obviously depend on the stochastic process that controls the statistics of Ω. Once we have calculated the factorial moments, any particular moment of the counting distribution can be obtained.

We will be content with the lowest two factorial moments; these give the mean and the variance of the photoelectron counting distribution. We have

$$n^{(1)} = \sum_{n=0}^{\infty} np(n,t,T) = \bar{n}, \tag{5.15}$$

where the overbar denotes the average against $p(n,t,T)$ (in contrast to the angular brackets that denote the average over the stochastic intensity). Using (5.14) the *mean of the photoelectron counting distribution* is given by

$$\bar{n} = \left\langle \xi \int_t^{t+T} dt' \bar{I}(t') \right\rangle. \tag{5.16}$$

This is exactly what we would expect. For a constant intensity it reduces to (5.3). The variance of the photoelectron counting distribution is defined by

$$\overline{\Delta n^2} = \overline{n^2} - \bar{n}^2. \tag{5.17}$$

In terms of factorial moments we have

$$\overline{\Delta n^2} = n^{(2)} + \bar{n} - \bar{n}^2. \tag{5.18}$$

Then from (5.14), we obtain

$$n^{(2)} = \xi^2 \int_t^{t+T} dt' \int_t^{t+T} dt'' \langle \bar{I}(t')\bar{I}(t'') \rangle. \tag{5.19}$$

Thus, the *variance of the photoelectron counting distribution* is given by

$$\overline{\Delta n^2} = \bar{n} + \xi^2 \int_t^{t+T} dt' \int_t^{t+T} dt'' \left[\langle \bar{I}(t')\bar{I}(t'') \rangle - \langle \bar{I}(t') \rangle \langle \bar{I}(t'') \rangle \right]. \tag{5.20}$$

The variance (5.20) has a number of properties that are important to note. First, it differs from the Poisson result $\overline{\Delta n^2} = \bar{n}$ for constant intensity light by an amount that depends on the intensity correlations. More precisely, we write

$$\langle \bar{I}(t')\bar{I}(t'') \rangle - \langle \bar{I}(t') \rangle \langle \bar{I}(t'') \rangle = \langle \bar{I}(t') \rangle \langle \bar{I}(t'') \rangle \left[g^{(2)}(t',t'') - 1 \right], \tag{5.21}$$

where

$$g^{(2)}(t',t'') \equiv \frac{\langle \bar{I}(t')\bar{I}(t'') \rangle}{\langle \bar{I}(t') \rangle \langle \bar{I}(t'') \rangle} \tag{5.22}$$

is the normalized *second-order correlation function* (or degree of second-order coherence). The deviation from a Poisson variance then depends on the second-order correlation function. An optical field is said to possess *second-order coherence* if

$$g^{(2)}(t',t'') = 1. \tag{5.23}$$

For such fields the photoelectron counting distribution is a Poisson distribution.

Secondly, within the confines of classical stochastics, the fluctuations in $\bar{I}(t)$ can only *broaden* the photoelectron counting distribution beyond that for a Poisson distribution, producing a *super-Poissonian distribution.* To illustrate this we first specialize (5.20) to the case of a stationary field. For stationary fields the origin of time is unimportant and we may write

$$\overline{\Delta n^2} = \bar{n} + \xi^2 \int_0^T dt' \int_0^T dt'' \langle \bar{I}(t')\bar{I}(t'')\rangle - \xi^2 T^2 \langle \bar{I}\rangle^2. \tag{5.24}$$

A more convenient form for the double integral is obtained by a further use of the stationary property, and a change of variables:

$$\begin{aligned}
&\int_0^T dt' \int_0^T dt'' \langle \bar{I}(t')\bar{I}(t'')\rangle \\
&\quad = \int_0^T dt' \int_0^{t'} dt'' \langle \bar{I}(t'-t'')\bar{I}(0)\rangle + \int_0^T dt' \int_{t'}^T dt'' \langle \bar{I}(0)\bar{I}(t''-t')\rangle \\
&\quad = \int_0^T dt' \int_0^{t'} d\tau \langle \bar{I}(\tau)\bar{I}(0)\rangle + \int_0^T dt' \int_0^{T-t'} d\tau \langle \bar{I}(0)\bar{I}(\tau)\rangle \\
&\quad = \int_0^T d\tau \int_\tau^T dt' \langle \bar{I}(\tau)\bar{I}(0)\rangle + \int_0^T d\tau \int_0^{T-\tau} dt' \langle \bar{I}(0)\bar{I}(\tau)\rangle \\
&\quad = 2\int_0^T d\tau (T-\tau)\langle \bar{I}(0)\bar{I}(\tau)\rangle.
\end{aligned}$$

Thus, for a stationary field, the deviation from a Poisson variance is given by

$$\overline{\Delta n^2} - \bar{n} = 2\xi^2 \int_0^T d\tau (T-\tau)\langle \bar{I}(0)\bar{I}(\tau)\rangle - \xi^2 T^2 \langle \bar{I}\rangle^2. \tag{5.25}$$

If we specialize to counting times that are short compared to the intensity correlation time of the light, we may set $\langle \bar{I}(0)\bar{I}(\tau)\rangle \approx \langle \bar{I}^2\rangle$; then it is easy to see that the right-hand side of (5.25) must be positive; we obtain

$$\overline{\Delta n^2} - \bar{n} = \xi^2 T^2 (\langle \bar{I}^2\rangle - \langle \bar{I}\rangle^2), \tag{5.26}$$

which is positive because the intensity variance on the right-hand side must be positive. A common measure of the deviation from the Poisson variance is *the Mandel Q parameter*:

$$Q \equiv \frac{\overline{\Delta n^2} - \bar{n}}{\bar{n}}, \tag{5.27}$$

which is generally a function of t and T. For all classical stochastic optical fields Q is nonnegative. In the case of a stationary field and a counting time much shorter than the intensity correlation time, (5.26) shows this explicitly:

$$Q = \xi T \frac{\langle \bar{I}^2 \rangle - \langle \bar{I} \rangle^2}{\langle \bar{I} \rangle}. \tag{5.28}$$

Thirdly, and finally, (5.28) indicates that the deviation from a Poisson distribution grows linearly with the counting time T. This is only true so long as T remains less than the intensity correlation time. For longer counting times, the photoelectric emissions that are separated by a large interval compared to the intensity correlation time are uncorrelated; they tend to move the counting distribution towards a Poisson distribution. But there remains an accumulated effect from those emissions that are separated by intervals smaller than the intensity correlation time. To illustrate the long counting time effects we use the example of (filtered) thermal light. The intensity correlation function is given by (3.87) ($\tau \geq 0$):

$$\langle \bar{I}(0)\bar{I}(\tau) \rangle = \langle \bar{I} \rangle^2 (1 + e^{-2\kappa\tau}). \tag{5.29}$$

The integral in (5.25) is readily evaluated and gives

$$\overline{\Delta n^2} - \bar{n} = \xi^2 \langle \bar{I} \rangle^2 \left(\frac{T}{\kappa} + \frac{e^{-2\kappa T} - 1}{2\kappa^2} \right), \tag{5.30a}$$

or,

$$Q = \frac{\xi \langle \bar{I} \rangle}{\kappa} \left(1 + \frac{e^{-2\kappa T} - 1}{2\kappa T} \right). \tag{5.30b}$$

This result agrees with (5.28) for $2\kappa T \ll 1$, since for thermal light $\langle \bar{I}^2 \rangle - \langle \bar{I} \rangle^2 = \langle \bar{I} \rangle^2$. For long counting times the deviation from a Poisson variance saturates at $Q = \xi \langle \bar{I} \rangle / \kappa$. Note that this is the mean number of photoelectrons emitted during an interval $-\tau_c$ to $+\tau_c$, where $\tau_c = (2\kappa)^{-1}$ is the intensity correlation time. Therefore, in a loose sense the Mandel Q parameter saturates at a value given by the mean number of (neighboring) photoelectrons that are correlated with an arbitrary photoelectron selected from a continuous sequence of photoelectric emissions.

5.4 The waiting-time distribution

In lecture 3 we mentioned the phenomenon of *photon bunching* for thermal light (Sect. 3.4). Now is perhaps a good time to discuss this phenomenon in a little more detail. It is usual to mention photon bunching in a discussion of the intensity correlation function, which is what we have done. But, actually, a better understanding of the phenomenon is gained by considering a related quantity – the distribution of waiting times between successive photoelectric emissions. We will call this distribution the *waiting-time distribution* and denote it by $w(\tau)$. It is defined as follows:

$$w(\tau)d\tau \equiv \begin{cases} \text{probability, given a photoelectric emission has just} \\ \text{ocurred, that there are no photoelectric emissions} \\ \text{during an interval of length } \tau\text{, followed by} \\ \text{a photoelectric emission during the next } d\tau. \end{cases} \tag{5.31}$$

We assume the process is stationary so that $w(\tau)$ is only a function of the waiting time τ. It is straightforward to evaluate this distribution for the random emission model considered in Sect. 5.1. We divide the waiting time into N subintervals of duration Δt. Then, in the limit $N \to \infty$, $\Delta t \to 0$, with $N\Delta t = \tau$ constant, from (5.5) and (5.6) we have

$$w(\tau)d\tau = (1 - \xi\bar{I}\Delta t)^N \xi\bar{I}\Delta t \to \xi\bar{I}\exp(-\xi\bar{I}\tau)d\tau. \tag{5.32}$$

The *mean waiting time* is given by

$$\bar{\tau} = \int_0^\infty d\tau\tau\xi\bar{I}e^{-\xi\bar{I}\tau} = (\xi\bar{I})^{-1}. \tag{5.33}$$

This is the ratio of the counting time T and the average number of photoelectric counts $\bar{n}$ $[T/\bar{n} = T/\xi\bar{I}T = (\xi\bar{I})^{-1}]$ which is what we would expect.

We might compare the exponential waiting-time distribution (5.32) with the second-order correlation function $g^{(2)}(\tau) = 1$ for random photoelectric emissions (constant intensity light). The second-order correlation function is proportional to the probability that a pair of photoelectrons are emitted, separated by a time τ. The probability is not conditioned on the requirement that no other photoelectrons be emitted during the interval τ. On the other hand, the waiting-time probability is conditioned in this way. The exponential decay simply indicates that it becomes more and more unlikely to see no photoelectric emissions during an interval τ as the length of the interval increases.

Now when the light intensity is a stochastic quantity the shape of the waiting-time distribution changes. The photoelectric emissions are no longer random; the photoelectrons produced at either end of a waiting time interval that is smaller than the intensity correlation time are correlated. We will discuss a method for calculating the changed waiting-time distribution in a future lecture. For the moment let us just motivate the change that occurs in a qualitative way. We consider the probabilities $p(1)$ and $p(2)$ for counting one and two photoelectrons, respectively, during a very short time $\Delta t \ll \tau_c$:

$$p(1) = \xi \langle \bar{I} \rangle \Delta t, \tag{5.34a}$$

$$p(2) = \xi^2 \langle \bar{I}^2 \rangle \Delta t^2. \tag{5.34b}$$

For deterministic, or coherent fields, $\langle \bar{I}^2 \rangle = \langle \bar{I} \rangle^2$, and therefore $p(2) = p(1)^2$ indicating that the two photoelectric emissions are independent. But for stochastic fields

$$p(2) = p(1)^2 + \xi^2 (\langle \bar{I}^2 \rangle - \langle \bar{I} \rangle^2) \Delta t^2. \tag{5.35}$$

The variance of the intensity fluctuations in the field gives an increased probability for a second photoelectric emission in the interval Δt compared with that obtained for coherent light *of the same intensity*. This feature is captured by the usual statement of photon bunching:

$$g^{(2)}(0) \equiv \frac{\langle \bar{I}^2 \rangle}{\langle \bar{I} \rangle^2} = \frac{\langle \bar{I}^2 \rangle - \langle \bar{I} \rangle^2}{\langle \bar{I} \rangle^2} \geq 1. \tag{5.36}$$

In fact, for intervals that are much less that the mean waiting time, the second-order correlation function $g^{(2)}(\Delta t)$ and the waiting-time distribution $w(\Delta t)$ are proportional to one another. This is because the probability for additional photoelectric emissions to occur *during* the interval becomes very small for $\Delta t \ll \bar{\tau}$. But, because it it a probability distribution, the waiting-time distribution has properties that allow us to extrapolate from the short-time behavior to a qualitative change in shape over all times. Although, for the stochastic field, there is an enhanced probability for two photoelectrons to be emitted in a short interval compared with the probability for coherent light of the same intensity, the average rate at which photoelectrons are emitted must be the same for the two fields (since they have the same intensity). We must imagine a redistribution of the waiting times that keeps the mean waiting time the same. The general character of this redistribution is illustrated in Fig. 5.2(a). The area under both of the curves in the figure is unity, and both have the same mean. To accomplish this the stochastic light must show an enhanced probability for short *and long* waiting times, and a decreased probability for intermediate waiting times. We can understand what this means by considering a random sequence of photoelectron emission times and asking how we must rearrange it to correspond to the changed waiting time distribution: As shown in Fig. 5.2(b), we must move some of the emission times to increase the number of clumps in the sequence, and also the number of gaps; thus, the photoelectron emission times are more bunched. The intensity correlation function $g^{(2)}(\tau) = 1 - e^{-2\kappa\tau}$ does not show this overall redistribution of emission times. It reproduces only the short-time behavior of $w(\tau)$.

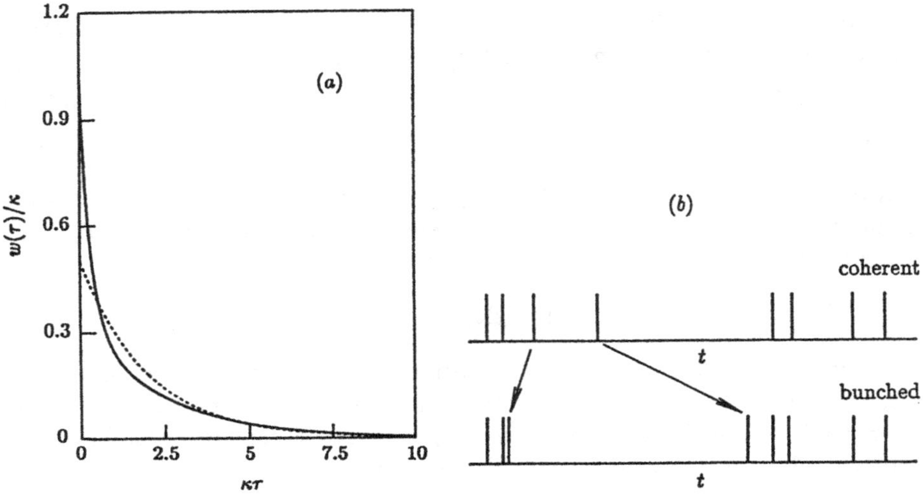

Fig. 5.2. (a) Comparison between the waiting-time distributions for filtered thermal light (bunched light) (solid curve) and coherent light of the same intensity (dashed curve). Broadband thermal light ($\bar{n} = 1$) is filtered by a cavity with linewidth κ (half-width at half-maximum) and equal transmission coefficients at the input and output mirrors. (The mean photon number in the cavity is $barn/2$.) (b) Rearrangement of a typical random photoelectron emission sequence to account for the change in the waiting-time distribution shown in (a).

5.5 Photoelectron counting for quantized fields

The general photoelectron counting distribution for classical fields is given in (5.11a) and (5.11b). We now want to know how this is changed for quantized fields. We might expect to make the replacement

$$\bar{I} \to 2\epsilon_0 c\hat{E}^{(-)}\hat{E}^{(+)}, \tag{5.37}$$

where $\hat{E}^{(+)}$ and $\hat{E}^{(-)}$ are the positive and negative frequency components of the electric field operator evaluated at the location of the detector, and the factor $2\epsilon_0 c$ is needed to give the units of intensity. The quantized field might, for example, be the output field from an optical cavity (Sect. 1.4), or the field radiated by a two-state atom (Sect. 2.4). With the replacement (5.37) we would interpret the average in (5.11a) as a quantum-mechanical average instead of an average over a classical stochastic intensity.

What we expect is essentially correct, but needs one small addition. Once we have operators we must face the issue of operator order. The appropriate order for the operators in the photoelectron counting distribution is normal order and time order. We illustrate this by the example of the intensity correlation function. For $t'' \geq t'$, the replacement is

$$\langle\bar{I}(t')\bar{I}(t'')\rangle \to (2\epsilon_0 c)^2\langle\hat{E}^{(-)}(t')\hat{E}^{(-)}(t'')\hat{E}^{(+)}(t'')\hat{E}^{(+)}(t')\rangle. \tag{5.38}$$

The operators are in *normal order* – all creation operators to the left and all annihilation operators to the right, and *time order* – time arguments increasing from the extreme left and right to take their largest values in the center. This is the operator ordering that appears in the correlation functions calculated in Sects. 3.4 and 3.5. The reasons for this order can be appreciated in general terms without too much effort. Insert unity as an expansion in a complete set of states in the middle of the average on the right-hand side of (5.38). Then we can see that this average is the sum (over final states) of squared probability amplitudes for the annihilation of two photons from the detected field; we see that the normal order arises because photoelectric detectors work by annihilating photons. The time ordering comes from the ordering of successive photon annihilations in a perturbative treatment of the interaction between the detector and the field. The details can be found in the work of Glauber on photoelectric detection and quantum coherence [5.3, 5.4].

If we can accept the operator ordering, having seen where the classical photoelectric counting distribution comes from we can essentially guess the form of the photoelectron counting distribution for quantized fields:

$$p(n,t,T) = \left\langle \, : \frac{\hat{\Omega}^n}{n!} e^{-\hat{\Omega}} : \right\rangle, \tag{5.39a}$$

where the integrated intensity is now an operator:

$$\hat{\Omega} = \xi \int_t^{t+T} dt' \hat{E}^{(-)}(t') \hat{E}^{(+)}(t'), \tag{5.39b}$$

with

$$\xi = \eta A \frac{2\epsilon_0 c}{\hbar\omega}. \tag{5.40}$$

The notation $\langle : : \rangle$ indicates that the operators are to be written in normal and time order. This, of course, cannot be done explicitly for something as complicated as the exponentiated integral in (5.39a). A thorough discussion of the theory of photoelectric counting for quantized fields, including the derivation of (5.39), is given by Kelly and Kleiner [5.5].

Our calculation of moments for classical fields carries through in an identical manner for quantized fields. In particular, the *mean of the photoelectron counting distribution* is

$$\bar{n} = \xi \int_t^{t+T} dt' \langle \hat{E}^{(-)}(t') \hat{E}^{(+)}(t') \rangle, \tag{5.41}$$

and the *variance of the photoelectron counting distribution* is given by

$$\begin{aligned}\overline{\Delta n^2} - \bar{n} \\ &= \xi^2 \int_t^{t+T} dt' \int_t^{t+T} dt'' \Big[\langle :\hat{E}^{(-)}(t')\hat{E}^{(+)}(t')\hat{E}^{(-)}(t'')\hat{E}^{(+)}(t''): \rangle \\ &\quad - \langle \hat{E}^{(-)}(t')\hat{E}^{(+)}(t')\rangle\langle \hat{E}^{(-)}(t'')\hat{E}^{(+)}(t'')\rangle\Big]. \end{aligned} \tag{5.42}$$

The argument of the double integral in (5.42) can be written as

$$\langle \hat{E}^{(-)}(t')\hat{E}^{(+)}(t')\rangle\langle \hat{E}^{(-)}(t'')\hat{E}^{(+)}(t'')\rangle \left[g^{(2)}(t',t'') - 1\right]$$

where

$$g^{(2)}(t',t'') = \frac{\langle :\hat{E}^{(-)}(t')\hat{E}^{(+)}(t')\hat{E}^{(-)}(t'')\hat{E}^{(+)}(t''): \rangle}{\langle \hat{E}^{(-)}(t')\hat{E}^{(+)}(t')\rangle\langle \hat{E}^{(-)}(t'')\hat{E}^{(+)}(t'')\rangle} \tag{5.43}$$

is the normalized *second-order correlation function for a quantized field.* What changes for the quantized fields are the properties of the averages – now operator averages. Corresponding to (5.25), for a stationary field we now have

$$\begin{aligned}\overline{\Delta n^2} - \bar{n} = 2\xi^2 \int_0^T d\tau (T-\tau)\langle \hat{E}^{(-)}(0)\hat{E}^{(-)}(\tau)\hat{E}^{(+)}(\tau)\hat{E}^{(+)}(0)\rangle \\ - \xi^2 T^2 \langle \hat{E}^{(-)}\hat{E}^{(+)}\rangle^2;\end{aligned} \tag{5.44}$$

for counting times much shorter than the intensity correlation time this gives

$$\begin{aligned}\overline{\Delta n^2} - \bar{n} &= \xi^2 T^2 \left(\langle \hat{E}^{(-)}\hat{E}^{(-)}\hat{E}^{(+)}\hat{E}^{(+)}\rangle - \langle \hat{E}^{(-)}\hat{E}^{(+)}\rangle^2\right) \\ &= \xi^2 T^2 \langle \hat{E}^{(-)}\hat{E}^{(+)}\rangle^2 \left[g^{(2)}(0) - 1\right].\end{aligned} \tag{5.45}$$

The quantum averages on the right-hand side of (5.45) are not constrained like the intensity variance on the right-hand side of (5.26) so that $\overline{\Delta n^2} - \bar{n}$ must be positive; we do not require $g^{(2)}(0) \geq 1$. Thus, it is possible for a quantized field to produce a ***sub-Poissonian*** photoelectron counting distribution. An example of this is provided by resonance fluorescence which has been seen to have $g^{(2)}(0) = 0$ (Sect. 3.5). The derivation of the photoelectron counting distribution for resonance fluorescence is rather involved and therefore we will not spend time on that here. Mollow provided the first derivation [5.6], and Cook developed an interesting indirect approach based on the theory of momentum transfer [5.7]. Mandel also did some early calculations and obtained results for the Q parameter (5.27) [5.8]; in subsequent experiments Short and Mandel observed the sub-Poissonian character of the photoelectron counting distribution [5.9]. A review of the work on resonance fluorescence is presented with a number of illustrations by Carmichael et al. [5.10]; one of the illustrations from this paper appears in Fig. 5.3.

In a related effect, the waiting times between photoelectrons in the detection of a quantized field can be distributed in a manner that is not possible

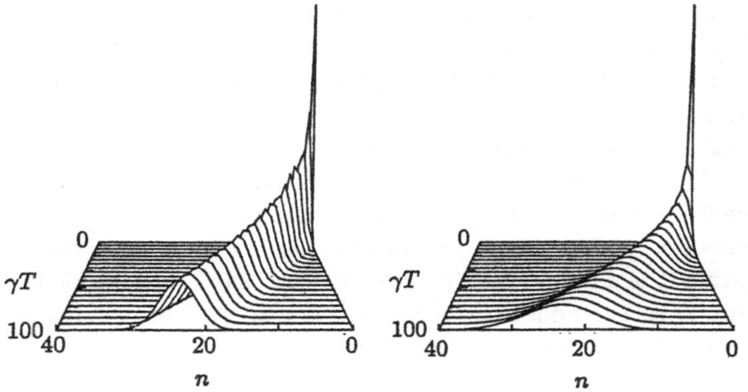

Fig. 5.3. Photoelectron counting distribution for resonance fluorescence (on the left) compared with the distribution for coherent light of the same intensity (on the right). The plot is for $\sqrt{2}\Omega/\gamma = 1$ and $\eta = 1$ where Ω is the Rabi frequency and γ is the Einstein A coefficient [the source master equation is (2.62)].

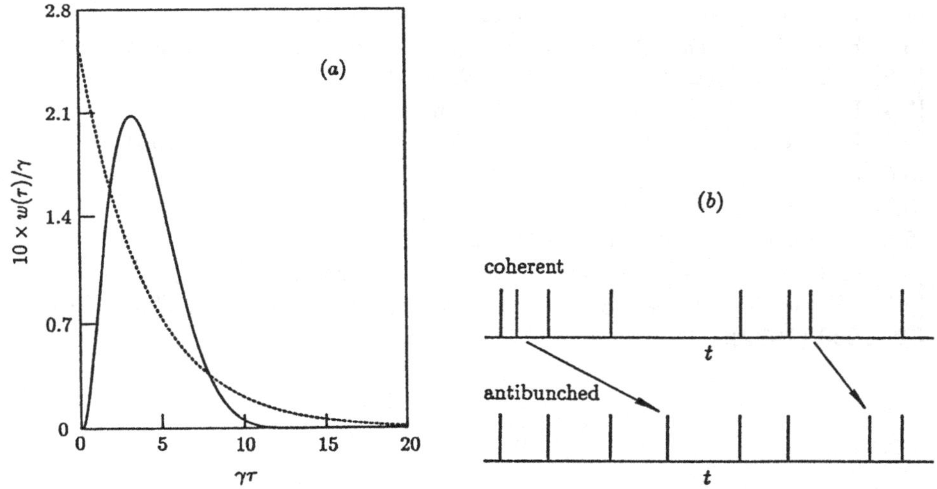

Fig. 5.4. (a) Comparison between the waiting-time distributions for resonance fluorescence (solid curve) and coherent light of the same intensity (dashed curve); parameter values are the same as in Fig. 5.3. (b) Rearrangement of a typical random photoelectron emission sequence to account for the change in the waiting-time distribution shown in (a).

for classical stochastic light. As we did before, consider the probabilities for detecting one and two photons in a short interval Δt:

$$p(1) = \xi\langle\hat{E}^{(-)}\hat{E}^{(+)}\rangle\Delta t, \tag{5.46a}$$

$$\begin{aligned} p(2) &= \xi^2\langle\hat{E}^{(-)}\hat{E}^{(-)}\hat{E}^{(+)}\hat{E}^{(+)}\rangle\Delta t^2 \\ &= p(1)^2 + \xi^2\langle\hat{E}^{(-)}\hat{E}^{(+)}\rangle^2\big[g^{(2)}(0) - 1\big]. \end{aligned} \tag{5.46b}$$

The term added to $p(1)^2$ can be negative for quantum fields, as illustrated by Fig. (3.2). The corresponding picture for the waiting-time distribution shows a decreased probability for short waiting times and long waiting times, and an increased probability for moderate waiting times (in comparison with coherent light of the same intensity). This is exactly the reverse of the situation illustrated in Fig. (5.2). An example of the waiting-time distribution for resonance fluorescence appears in Fig. 5.4(a), with the corresponding rearrangement of a typical random photoelectron emission sequence in Fig. 5.4(b). Together the figures clearly illustrate the *antibunching* of the photoelectron emissions; the emissions are more regular than a random sequence, tending towards an equal spacing in time.

References

[5.1] L. Mandel, *Proc. Phys. Soc.* **72**, 1037 (1958); *Progress in Optics*, Vol. 2, ed. by E. Wolf, North Holland: Amsterdam, 1963, pp. 181ff.
[5.2] R. Loudon, *The Quantum Theory of Light*, Oxford (1983), pp. 230ff.
[5.3] R. J. Glauber, *Phys. Rev.* **130**, 2529 (1963).
[5.4] R. J. Glauber, *Phys. Rev.* **131**, 2766 (1963).
[5.5] P. L. Kelly and W. H. Kleiner, *Phys. Rev.* **136**, A316 (1964).
[5.6] B. R. Mollow, *Phys. Rev.* A **12**, 1919 (1975).
[5.7] R. J. Cook, *Phys. Rev.* A **23**, 1243 (1981).
[5.8] L. Mandel, *Opt. Lett.* **4**, 205 (1979).
[5.9] R. Short and L. Mandel, *Phys. Rev. Lett.* **51**, 384 (1983).
[5.10] H. J. Carmichael, S. Singh, R. Vyas, and P. R. Rice, *Phys. Rev.* A **39**, 1200 (1989).

Lecture 6 – Photoelectric Detection II

In the last lecture we met some of the basic ideas behind photoelectric detection and photoelectron counting. We saw how one feature of the photoelectron counting distribution – its variance – distinguishes between optical fields described by a classical stochastic intensity and quantized fields. Quantized fields can produce a sub-Poissonian counting distribution; stochastic classical fields can only broaden the Poisson distribution obtained for constant intensity light. Similar distinctions between classical stochastic fields and quantized fields can show up in other ways. The photoelectron counting distribution looks at fluctuations in intensity. By using a homodyne technique we can use photoelectric counting to observe fluctuations in the field amplitude; the amplitude fluctuations of a quantized field can also do things that are not reproducible by classical stochastics – so-called *squeezing below the vacuum limit.* Homodyne detection and squeezing are the subjects of this lecture. We begin with a brief description of squeezed light.

6.1 Squeezed light

There are now many treatments of squeezing in the literature [6.1–6.4]. One convenient way to introduce squeezing is to analyze a simple physical system that generates squeezed light. This is the approach we will take. The simple system is the degenerate parametric amplifier, and before we start thinking about quantum fluctuations it is helpful to understand the phase-sensitive nature of this device using a classical theory.

We consider the classical theory of degenerate parametric amplification without pump depletion. The basic system is the lossless cavity illustrated in Fig. 6.1. The cavity supports two resonant electromagnetic field modes that couple through the $\chi^{(2)}_{yxy}$ $(= \chi^{(2)}_{yyx} = \chi^{(2)}_{xyy})$ component of the nonlinear susceptibility tensor of an intracavity crystal (for example, $LiNbO_3$ with the optic axis aligned in the x direction). These are the subharmonic and pump fields given, respectively, by

$$\boldsymbol{E}(z,t) = \hat{e}_y \hat{\mathcal{E}}(t) A(z) \cos[\Phi(z) + \phi] e^{-i\omega_C t} + \text{c.c.}, \tag{6.1a}$$

$$\boldsymbol{E}_p(z,t) = \hat{e}_x \hat{\mathcal{E}}_p A(z) \cos[2\Phi(z) + \phi_p] e^{-i2\omega_C t} + \text{c.c.}, \tag{6.1b}$$

with

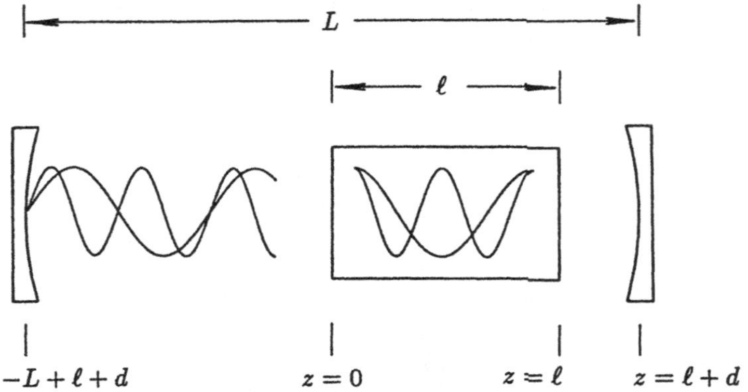

Fig. 6.1. Cavity geometry for a standing-wave degenerate parametric amplifier. The relative phases of the standing-wave pump and subharmonic mode functions are shown inside the crystal for maximum coupling, and as determined by reflection boundary conditions at the mirrors. The incompatible standing-wave patterns must be matched with the use of dispersive elements inside the cavity.

$$A(z) = 1 + \left(1/\sqrt{n} - 1\right)[\theta(z) - \theta(z - \ell)], \tag{6.2a}$$

$$\Phi(z) = (\omega_C/c)z + (n-1)(\omega_C/c)\int_0^z dz'[\theta(z') - \theta(z' - \ell)], \tag{6.2b}$$

where $\theta(\xi) = 0$ for $\xi < 0$, and $\theta(\xi) = 1$ for $\xi \geq 1$; $\hat{\mathcal{E}}(t)$ and $\hat{\mathcal{E}}_p$ are complex mode amplitudes, ϕ and ϕ_p are constants that determine the phases of the standing waves in the crystal, $\hat{e}_x$ and $\hat{e}_y$ are unit polarization vectors, and n is the crystal refractive index. We assume perfect phase matching, and small parametric gain so that the forwards and backwards field amplitudes in (6.1) can be taken equal. In the undepleted pump approximation we take $\hat{\mathcal{E}}_p$ to be constant; we seek an equation of motion for the amplitude $\hat{\mathcal{E}}(t)$.

The interaction between modes inside the crystal creates the polarization

$$\boldsymbol{P}_{\omega_C}(z,t) = \hat{e}_y(2\epsilon_0\chi^{(2)}/n)\hat{\mathcal{E}}^*(t)\hat{\mathcal{E}}_p \cos[\Phi(z) + \phi]\cos[2\Phi(z) + \phi_p]e^{-i\omega_C t} + \text{c.c.}, \tag{6.3}$$

oscillating at the frequency ω_C, where $\chi^{(2)} \equiv \chi^{(2)}_{yxy} = \chi^{(2)}_{yyx} = \chi^{(2)}_{xyy}$. More precisely, we identify the polarization components that radiate the forwards and backwards traveling subharmonic waves by expanding the product of cosines in (6.3) as a sum of exponentials; thus, forwards and backwards waves

$$\boldsymbol{E}_f(z,t) = \hat{e}_y\tfrac{1}{2}\hat{\mathcal{E}}(t)A(z)\exp\left[-i\left(\omega_C t - \Phi(z) - \phi\right)\right], \tag{6.4a}$$

and

$$\boldsymbol{E}_b(z,t) = \hat{e}_y\tfrac{1}{2}\hat{\mathcal{E}}(t)A(z)\exp\left[-i\left(\omega_C t + \Phi(z) + \phi\right)\right], \tag{6.4b}$$

respectively, are radiated by $(0 < z < \ell)$

$$\boldsymbol{P}^f_{\omega_C}(z,t) = \hat{e}_y \tfrac{1}{2}\mathcal{P}_f(t)e^{-i[\omega_C(t-z/c)-\phi]} + \text{c.c.}, \tag{6.5a}$$
$$\mathcal{P}_f(t) = (\epsilon_0\chi^{(2)}/n)\hat{\mathcal{E}}^*(t)\hat{\mathcal{E}}_p e^{i(\phi_p-2\phi)}, \tag{6.5b}$$

and

$$\boldsymbol{P}^b_{\omega_C}(z,t) = \hat{e}_y \tfrac{1}{2}\mathcal{P}_b(t)e^{-i[\omega_C(t+z/c)+\phi]} + \text{c.c.}, \tag{6.6a}$$
$$\mathcal{P}_b(t) = (\epsilon_0\chi^{(2)}/n)\hat{\mathcal{E}}^*(t)\hat{\mathcal{E}}_p e^{-i(\phi_p-2\phi)}. \tag{6.6b}$$

When the parametric gain is small the subharmonic field amplitude only changes significantly after making many round trips in the cavity. Its rate of change can be obtained from the ratio of the change $\Delta\hat{\mathcal{E}}$ on a single round trip and the cavity round-trip time

$$t_C = 2\bar{L}/c, \qquad \bar{L} \equiv L + (n-1)\ell; \tag{6.7}$$

L is the cavity length and ℓ is the length of the crystal. By following the forwards field at $z=0$ once around the cavity, we find

$$\begin{aligned}\hat{\mathcal{E}} + \Delta\hat{\mathcal{E}} = &\left\{\left(\frac{1}{\sqrt{n}}\hat{\mathcal{E}} + i\frac{\omega_C\ell}{2\epsilon_0 cn}\mathcal{P}_f\right)\sqrt{n}e^{i\phi_R}e^{2i[\Phi(\ell+d)+\phi]}\frac{1}{\sqrt{n}} + i\frac{\omega_C\ell}{2\epsilon_0 cn}\mathcal{P}_b\right\}\\ &\times\sqrt{n}e^{i\phi_R}e^{-2i[\Phi(-L+\ell+d)+\phi]},\end{aligned} \tag{6.8}$$

where the terms $i(\omega_C\ell/2\epsilon_0 cn)\mathcal{P}_f$ and $i(\omega_C\ell/2\epsilon_0 cn)\mathcal{P}_b$ are the increments added to the field amplitude due to forwards and backwards propagation, respectively, through the crystal; the factors $1/\sqrt{n}$ and $\sqrt{n}$ transform field amplitudes into and out of the crystal, ϕ_R is a phase change due to reflection at the mirrors, and $2[\Phi(\ell+d)+\phi]$ and $-2[\Phi(-L+\ell+d)+\phi]$ are phase changes required by boundary conditions at the mirrors. Substituting (6.5b) and (6.6b) into (6.7), and using the resonance condition

$$2[\phi_R + \Phi(\ell+d) - \Phi(-L+\ell+d)] = N2\pi, \qquad N \text{ an integer}, \tag{6.9}$$

and the boundary condition at $z = \ell + d$,

$$\phi_R + 2[\Phi(\ell+d)+\phi] = M2\pi, \qquad M \text{ an integer}, \tag{6.10}$$

we obtain the equation of motion for the subharmonic field amplitude:

$$\dot{\hat{\mathcal{E}}} = \frac{\Delta\hat{\mathcal{E}}}{2\bar{L}/c} = K\hat{\mathcal{E}}^*, \tag{6.11}$$

with

$$K = i\frac{\omega_C\ell\chi^{(2)}}{2n^{3/2}\bar{L}}\hat{\mathcal{E}}_p\cos(\phi_p - 2\phi). \tag{6.12}$$

The solution to (6.11) is best expressed in terms of quadrature phase amplitudes of the subharmonic field. For an arbitrary choice of phase θ, *quadrature phase amplitudes* $\hat{\mathcal{E}}_\theta(t)$ and $\hat{\mathcal{E}}_{\theta+\pi/2}(t)$ are defined by writing (6.1a) in the form

$$\boldsymbol{E}(z,t) = \hat{e}_y 2\cos[\Phi(z)+\phi]\big[\hat{\mathcal{E}}_\theta(t)\cos(\omega_C t-\theta) + \hat{\mathcal{E}}_{\theta+\pi/2}(t)\sin(\omega_C t-\theta)\big], \tag{6.13}$$

with

$$\hat{\mathcal{E}}_\theta = \tfrac{1}{2}\big(\hat{\mathcal{E}}e^{-i\theta} + \hat{\mathcal{E}}^* e^{i\theta}\big). \tag{6.14}$$

Equation (6.11) is then equivalent to the pair of equations

$$\dot{\mathcal{X}} = |K|\mathcal{X}, \qquad \dot{\mathcal{Y}} = -|K|\mathcal{Y}, \tag{6.15}$$

where $\mathcal{X} \equiv \hat{\mathcal{E}}_\theta$, $\mathcal{Y} \equiv \hat{\mathcal{E}}_{\theta+\pi/2}$, with $\theta = \frac{1}{2}\arg(K)$. The solution to these equations gives

$$\mathcal{X}(t) = e^{|K|t}\mathcal{X}(0), \qquad \mathcal{Y}(t) = e^{-|K|t}\mathcal{Y}(0). \tag{6.16}$$

Thus, the degenerate parametric amplifier is a phase-sensitive amplifier; with the appropriate choice of phase, one quadrature phase amplitude of the subharmonic field is amplified and the other is deamplified.

Let us now convert our model into a quantum-mechanical form. In the language of quantum mechanics the energy exchange in degenerate parametric amplification results from an interaction that annihilates one pump photon with frequency $2\omega_C$, and creates two subharmonic photons with frequency ω_C. The conjugate interaction describes the process of second harmonic generation. In the undepleted pump approximation the pump mode is assumed to be highly populated, and the loss or gain of photons in this mode is assumed to be negligible. The pump is then treated as a classical field with constant amplitude. The Hamiltonian describing the creation and annihilation of photons in the subharmonic mode is given by

$$H = \hbar\omega_C a^\dagger a + i\hbar\tfrac{1}{2}\big(Ke^{-i2\omega_C t}a^{\dagger 2} - K^* e^{i2\omega_C t}a^2\big), \tag{6.17}$$

where K is the coupling constant defined by (6.12). From this Hamiltonian we obtain the Heisenberg equations of motion

$$\dot{\tilde{a}} = K\tilde{a}^\dagger, \qquad \dot{\tilde{a}}^\dagger = K^*\tilde{a}; \tag{6.18}$$

$\tilde{a}$ and $\tilde{a}^\dagger$ are annihilation and creation operators in a frame rotating at the frequency ω_C [Eq. (1.53)]. Equations (6.18) are the quantized version of equations (6.11). To complete the translation into quantum-mechanical language we must replace the classical field (6.1a) by the operator

$$\hat{\boldsymbol{E}}(z,t) = i\hat{e}_y\sqrt{\frac{\hbar\omega_C}{\epsilon_0 A\bar{L}}}\,A(z)\cos[\Phi(z)+\phi]\Big(\tilde{a}(t)e^{-i\omega_C t} - \tilde{a}^\dagger(t)e^{i\omega_C t}\Big), \tag{6.19}$$

and define operator quadrature phase amplitudes

$$\hat{A}_\theta = \tfrac{1}{2}\big(\tilde{a}e^{-i\theta} + \tilde{a}^\dagger e^{i\theta}\big). \tag{6.20}$$

The amplification and deamplification of quadrature phase amplitudes occurs in much the same way in the quantum theory as it does in the classical theory. We write the Heisenberg equations of motion (6.18) as

$$\dot{\hat{X}} = |K|\hat{X}, \qquad \dot{\hat{Y}} = -|K|\hat{Y}, \tag{6.21}$$

where $\hat{X} \equiv \hat{A}_\theta$, $\hat{Y} \equiv \hat{A}_{\theta+\pi/2}$, with $\theta = \frac{1}{2}\arg(K)$. Then

$$\hat{X}(t) = e^{|K|t}\hat{X}(0), \qquad \hat{Y}(t) = e^{-|K|t}\hat{Y}(0). \tag{6.22}$$

There is one important difference, however. $\hat{X}$ and $\hat{Y}$ are now operators; the quadrature phase amplitudes exhibit fluctuations

$$\Delta A_\theta = \sqrt{\langle(\hat{A}_\theta - \langle\hat{A}_\theta\rangle)^2\rangle}. \tag{6.23}$$

The size of these fluctuations will depend on the state of the field. In principle, in any one quadrature phase amplitude the fluctuations may be arbitrarily small. But, according to the Heisenberg uncertainty principle, for any θ, the product $\Delta A_\theta \Delta A_{\theta+\pi/2}$ is bounded below, with

$$\Delta A_\theta \Delta A_{\theta+\pi/2} \geq \tfrac{1}{2}|\langle[\hat{A}_\theta, \hat{A}_{\theta+\pi/2}]\rangle| = \tfrac{1}{4} \tag{6.24}$$

This uncertainty relation is equivalent to the uncertainty relation satisfied by the position and momentum operators of a mechanical oscillator.

A freely evolving field mode prepared in a coherent state satisfies

$$\Delta A_\theta \Delta A_{\theta+\pi/2} = \tfrac{1}{4}, \text{ with } \Delta A_\theta = \Delta_{\theta+\pi/2} = \tfrac{1}{2}. \tag{6.25}$$

The picture of the quantum fluctuations drawn from these results is illustrated in Fig. 6.2(a). What happens to a coherent field when it is amplified by a degenerate parametric amplifier? Of course, the mean quadrature phase amplitudes $\langle\hat{X}\rangle$ and $\langle\hat{Y}\rangle$ are respectively amplified and deamplified like the quadrature phase amplitudes $\mathcal{X}$ and $\mathcal{Y}$ in the classical theory. But what about the quantum fluctuations? Equations (6.22), (6.23) and (6.25) provide the simple answer: from these equations,

$$\Delta X(t) = e^{|K|t}\Delta X(0) = e^{|K|t}\tfrac{1}{2}, \tag{6.26a}$$

$$\Delta Y(t) = e^{-|K|t}\Delta Y(0) = e^{-|K|t}\tfrac{1}{2}. \tag{6.26b}$$

Thus, the fluctuations in quadrature phase amplitudes [with $\theta = \frac{1}{2}\arg(K)$] are amplified and deamplified in the same fashion as the means, and continue to satisfy the minimum uncertainty requirement $\Delta X \Delta Y = \frac{1}{4}$. This amplification and deamplification of the quantum fluctuations is illustrated for an initial vacuum state in Fig. 6.2(b). More generally, for an initial coherent state $|\alpha\rangle$, the ellipse in Fig. 6.2(b) is displaced from the origin to the point $X = \exp(|K|t)|\alpha|\cos[\arg(\alpha) - \frac{1}{2}\arg(K)]$, $Y = \exp(-|K|t)|\alpha|\sin[\arg(\alpha) - \frac{1}{2}\arg(K)]$.

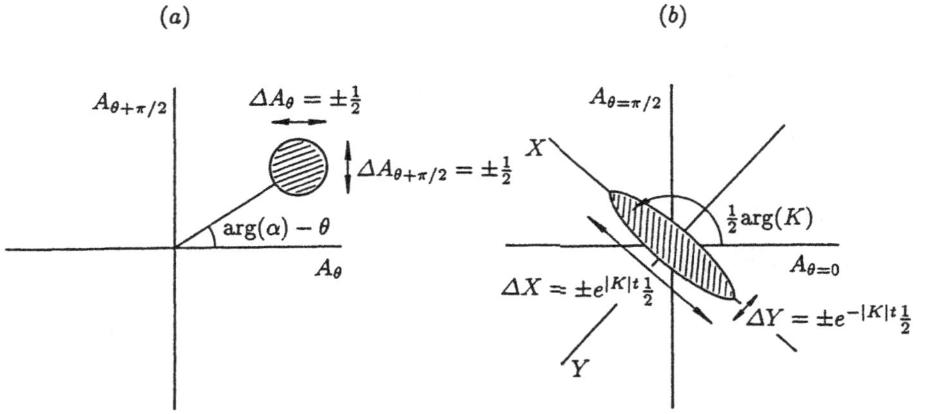

Fig. 6.2. Phase-space picture of the quantum fluctuations in (a) a freely evolving field mode prepared in the coherent state $|\alpha\rangle$, and (b) the subharmonic field of a degenerate parametric amplifier prepared in the vacuum state. Fluctuations in the field amplitude explore the shaded regions of phase space. For a rigorous interpretation A_θ must be read as a quadrature phase amplitude defined in terms of the complex argument α of the Wigner distribution [Eq. (6.34b)]. In a nonrotating frame the circle and the ellipse rotate clockwise about the origin at the frequency ω_C; the ordinate is then proportional to the oscillating electric field.

We now change our viewpoint, from the Heisenberg picture to the Schrödinger picture. In the Schrödinger picture we see that the degenerate parametric amplifier changes an initial coherent state of the subharmonic field into a squeezed coherent state. From the Hamiltonian (6.17), Schrödinger's equation in the interaction picture reads

$$\frac{d|\tilde{\psi}(t)\rangle}{dt} = \tfrac{1}{2}\left(K a^{\dagger 2} - K^* a^2\right)|\tilde{\psi}(t)\rangle, \tag{6.27}$$

where $|\tilde{\psi}(t)\rangle = e^{i\omega_C a^\dagger a t}|\psi(t)\rangle$, and we have used (1.24a). For an initial coherent state $|\alpha_0\rangle$,

$$\begin{aligned}|\psi(t)\rangle &= e^{-i\omega_C a^\dagger a t}\exp\left[\tfrac{1}{2}\left(K a^{\dagger 2} - K^* a^2\right)t\right]|\alpha_0\rangle \\ &= S\left(e^{i\pi}e^{-i2\omega_C t}Kt\right)|e^{-i\omega_C t}\alpha_0\rangle,\end{aligned} \tag{6.28}$$

where

$$S(\xi) = \exp\left[\tfrac{1}{2}(\xi^* a^2 - \xi a^{\dagger 2})\right]. \tag{6.29}$$

The unitary operator $S(\xi)$ is known as the *squeeze operator*, and the *squeezed coherent states* are defined by

$$|\alpha,\xi\rangle = D(\alpha)S(\xi)|0\rangle. \tag{6.30}$$

$S(\xi)$, $\xi = re^{i2\theta}$, squeezes the vacuum state to produce a squeezed vacuum state with $\Delta A_\theta = \frac{1}{2}e^{-r}$, $\Delta A_{\theta+\pi/2} = \frac{1}{2}e^{r}$, where $\hat{A}_\theta$ and $\hat{A}_{\theta+\pi/2}$ are defined by (6.20) (with the tilde removed). The *displacement operator*

$$D(\alpha) = \exp(\alpha a^\dagger - \alpha^* a) \tag{6.31}$$

adds the coherent amplitude α. In (6.28) the squeeze operator acts on the initial coherent amplitude as well as the fluctuations. With this taken into account we may write (6.28) as

$$|\psi(t)\rangle = |\alpha(t), \xi(t)\rangle, \tag{6.32}$$

where [with $\theta = \pi/2 - \omega_C t + \frac{1}{2}\arg(K)$]

$$\begin{aligned}\alpha(t) &= e^{i\theta}\Big\{e^{-|K|t}\tfrac{1}{2}\big[\alpha_0 e^{-i(\omega_C t+\theta)} + \alpha_0^* e^{i(\omega_C t+\theta)}\big] \\ &\qquad + e^{|K|t}\tfrac{1}{2}\big[\alpha_0 e^{-i(\omega_C t+\theta)} - \alpha_0^* e^{i(\omega_C t+\theta)}\big]\Big\} \\ &= e^{-i\omega_C t}\big[\alpha_0 \cosh(|K|t) + \alpha_0^* e^{i\arg(K)}\sinh(|K|t)\big],\end{aligned} \tag{6.33a}$$

and

$$\xi(t) = e^{i\pi}e^{-i2\omega_C t}Kt. \tag{6.33b}$$

It is helpful to picture a squeezed state using the quantum-classical correspondence discussed in Sect. 4.1. The Glauber-Sudarshan P distribution for a squeezed state does not exist as a well-behaved function; to represent squeezed states in this representation we have to use generalized functions [see the discussion below (4.10)]. However, Q and Wigner distributions do exist. For a squeezed vacuum state these are given by

$$\begin{aligned}Q(x+iy, x-iy) &= \sqrt{\frac{2}{\pi(1+e^{-2r})}}\exp\left[-\frac{1}{2}\frac{(x\cos\theta + y\sin\theta)^2}{(1+e^{-2r})/4}\right] \\ &\quad \times \sqrt{\frac{2}{\pi(1+e^{2r})}}\exp\left[-\frac{1}{2}\frac{(-x\sin\theta + y\cos\theta)^2}{(1+e^{2r})/4}\right],\end{aligned} \tag{6.34a}$$

and

$$\begin{aligned}W(x+iy, x-iy) &= \sqrt{\frac{2}{\pi e^{-2r}}}\exp\left[-\frac{1}{2}\frac{(x\cos\theta + y\sin\theta)^2}{e^{-2r}/4}\right] \\ &\quad \times \sqrt{\frac{2}{\pi e^{2r}}}\exp\left[-\frac{1}{2}\frac{(-x\sin\theta + y\cos\theta)^2}{e^{2r}/4}\right].\end{aligned} \tag{6.34b}$$

Note the larger variance in Q compared to W [see the discussion below (4.35)]. The variance of $\hat{A}_\theta$ involves the symmetrically-ordered product $\frac{1}{2}(a^\dagger a + a a^\dagger)$; it is for this reason that the Gaussian widths of the Wigner distribution match the standard deviations ΔA_θ and $\Delta A_{\theta+\pi/2}$ displayed in Fig. 6.2.

6.2 Homodyne detection: the spectrum of squeezing

For the rest of the lecture our source of squeezed light is a degenerate parametric oscillator. The parametric amplifier becomes a parametric oscillator when we include cavity loss and a pump field injected from outside the cavity. The master equation for this source is given in (2.63). Parametric amplifiers and oscillators are related in the same way as inversion based optical amplifiers and laser oscillators. Like a laser, a parametric oscillator has a threshold. The degenerate parametric oscillator produces squeezed light when it is operated below threshold [6.5].

We now turn to the primary interest of this lecture. What version of photoelectric detection can we use to observe squeezed light? Since squeezing is a phase-dependent phenomenon, clearly the scheme we choose must introduce a phase reference. Homodyne detection is then a natural candidate. In homodyne detection a strong local oscillator field is added to the field to be measured (the signal field). We therefore consider measurements made on the field described by the operator

$$\hat{\mathcal{E}}(t) = e^{-i\omega_C t}[\hat{\mathcal{E}}_{lo} + \xi\Delta\tilde{a}(t')], \tag{6.35}$$

where $\langle\hat{\mathcal{E}}_{lo}\rangle = \hat{\mathcal{E}}_{lo} = |\hat{\mathcal{E}}_{lo}|e^{i\theta}$ is the coherent local oscillator amplitude, t' is a retarded time [as in (1.60), for example], $\Delta\tilde{a}(t)$ describes fluctuations in the subharmonic mode – $\Delta\tilde{a} = \tilde{a} - \langle\tilde{a}\rangle$ – and ξ scales the subharmonic field so that $\xi\Delta\tilde{a}(t)$ has photon flux units; $\hat{\mathcal{E}}(t)$ and $\hat{\mathcal{E}}_{lo}$ also have photon flux units. If $\langle\tilde{a}\rangle \neq 0$ we can regard the nonzero mean to be included in the local oscillator amplitude. Now the probability of a photoelectric emission being produced by the combined field depends on the intensity $(\hat{\mathcal{E}}^{\dagger}\hat{\mathcal{E}})(t)$, and is sensitive to the relative phase between the local oscillator amplitude and the squeezed fluctuations in the subharmonic field. More precisely, the photoelectron counting distribution for an ideal detector (unit detection efficiency) and a counting interval $(t - \Delta t, t]$ is given by

$$p(n, t, \Delta t) = \left\langle : \frac{[(\hat{\mathcal{E}}^{\dagger}\hat{\mathcal{E}})(t)]^n \Delta t^n}{n!} \exp[-(\hat{\mathcal{E}}^{\dagger}\hat{\mathcal{E}})(t)\Delta t] : \right\rangle, \tag{6.36}$$

where we have assumed that Δt is much less than the correlation time when evaluating the integral (5.39b).

While we could define the spectrum of squeezing directly in terms of a photoelectric counting distribution, in practice the high photon flux associated with the strong local oscillator intensity makes photoelectric counting inappropriate. Instead we define the spectrum of squeezing in terms of the fluctuations of an analogue current. We must therefore say something about the way in which the intensity operator $(\hat{\mathcal{E}}^{\dagger}\hat{\mathcal{E}})(t)$ is turned into an electric current $i(t)$. The following analysis is based on the treatment by Carmichael [6.6].

Let us assume that a single photoelectric detection event produces a current pulse of width τ_d and amplitude Ge/τ_d, as illustrated in Fig. 6.3(a); e is the electronic charge and G is the gain. Then Fig. 6.3(b) shows how the photocurrent

$$i(t) = n_t \frac{Ge}{\tau_d} \tag{6.37}$$

is formed from the overlap of the n_t current pulses initiated during the interval $t - \tau_d$ to t; $i(t)$ is a classical stochastic process and n_t is a random variable. We can now use the photoelectric counting formula (6.36) to relate the classical fluctuations in $i(t)$ to the quantum fluctuations in the detected field. To derive the spectrum of photocurrent fluctuations we will need the autocorrelation function $\overline{i(t)i(t+\tau)}$. However, to see how things work, it is easier first to calculate the variance

$$\begin{aligned}
&\overline{i(t)i(t)} - \left(\overline{i(t)}\right)^2 \\
&= \left(\frac{Ge}{\tau_d}\right)^2 \left[\sum_n n^2 p(n,t,\tau_d) - \left(\sum_n np(n,t,\tau_d)\right)^2\right] \\
&= \left(\frac{Ge}{\tau_d}\right)^2 \left[\langle \hat{\mathcal{E}}^\dagger(t)\hat{\mathcal{E}}^\dagger(t)\hat{\mathcal{E}}(t)\hat{\mathcal{E}}(t)\rangle\tau_d^2 + \langle\hat{\mathcal{E}}^\dagger(t)\hat{\mathcal{E}}(t)\rangle\tau_d - \langle\hat{\mathcal{E}}^\dagger(t)\hat{\mathcal{E}}(t)\rangle^2\tau_d^2\right].
\end{aligned} \tag{6.38}$$

After substituting the field operator from (6.35) and taking the strong local oscillator limit, we find

$$\begin{aligned}
&\overline{i(t)i(t)} - \left(\overline{i(t)}\right)^2 \\
&= (Ge)^2\Big\{|\hat{\mathcal{E}}_{lo}|^4 + |\hat{\mathcal{E}}_{lo}|^2\xi^2\Big[4\langle\Delta\tilde{a}^\dagger(t)\Delta\tilde{a}(t)\rangle + e^{-2i\theta}\langle\Delta\tilde{a}(t)\Delta\tilde{a}(t)\rangle \\
&\quad + e^{2i\theta}\langle\Delta\tilde{a}^\dagger(t)\Delta\tilde{a}^\dagger(t)\rangle\Big] + |\hat{\mathcal{E}}_{lo}|^2\tau_d^{-1} - \Big[|\hat{\mathcal{E}}_{lo}|^4 + 2|\hat{\mathcal{E}}_{lo}|^2\xi^2\langle\Delta\tilde{a}^\dagger(t)\Delta\tilde{a}(t)\rangle\Big]\Big\} \\
&= (Ge)^2|\hat{\mathcal{E}}_{lo}|^2\xi^2 4\langle:\Delta\hat{A}_\theta(t)\Delta\hat{A}_\theta(t):\rangle + (Ge)^2|\hat{\mathcal{E}}_{lo}|^2\tau_d^{-1};
\end{aligned} \tag{6.39}$$

$\Delta\hat{A}_\theta = \hat{A}_\theta - \langle\hat{A}_\theta\rangle$ where $\hat{A}_\theta$ is defined in (6.20).

For a given local oscillator phase θ, the noise in the photocurrent depends on the field fluctuations described by the quadrature phase operator $\Delta\hat{A}_\theta$; different quadrature phase amplitudes of the subharmonic field can be selected by varying θ. Now, how do we set the level of squeezing in a quantitative fashion? The point of reference is set by considering what, in classical language, is a noiseless signal field. Assume the subharmonic mode is in a coherent state; it might as well be the vacuum state. Then the average $\langle:\Delta\hat{A}_\theta(t)\Delta\hat{A}_\theta(t):\rangle$ vanishes and the photocurrent fluctuates with variance

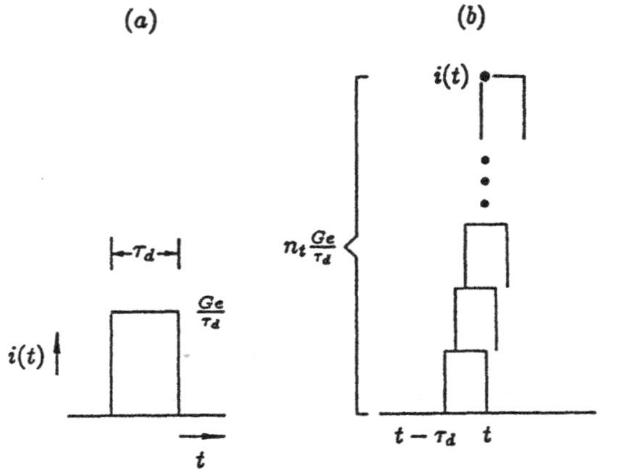

Fig. 6.3. (a) Current pulse produced by a single photodetection event. (b) Construction of the instantaneous photocurrent $i(t)$ from the overlap of n_t current pulses initiated during the interval $t - \tau_d$ to t.

$(Ge)^2|\hat{\mathcal{E}}_{lo}|^2\tau_d^{-1}$. This is the *shot noise* associated with the detection of the local oscillator intensity $|\hat{\mathcal{E}}_{lo}|^2$ – the Poisson variance derived in Sect. (5.1). Squeezed light has $\langle : \Delta\hat{A}_\theta(t)\Delta\hat{A}_\theta(t) : \rangle < 0$, which reduces the photocurrent fluctuations below this shot noise (vacuum state) level. The level of squeezing is defined by the size of the photocurrent variance relative to the shot noise level. However, we do not simple take the ratio of the two terms in (6.39). This ratio depends on τ_d, and the shot noise always dominates in the limit $\tau_d \to 0$. The reason is that the photocurrent variance is the integral, over all frequencies, of the power spectrum

$$P_\theta(\omega) = \frac{1}{\pi}\int_0^\infty d\tau \cos\omega\tau \lim_{t\to\infty}\left[\overline{i(t)i(t+\tau)} - \left(\overline{i(t)}\right)^2\right]. \tag{6.40}$$

Thus, the shot noise term in (6.39) corresponds to the frequency-space noise level $(Ge)^2|\hat{\mathcal{E}}_{lo}|^2/2\pi$ per unit bandwidth, multiplied by a bandwidth $2\pi/\tau_d$. The bandwidth is infinite when $\tau_d \to 0$.

To define the spectrum of squeezing we compare the contributions to the photocurrent fluctuations in frequency space. In the limit $\tau_d \to 0$ the correlation function needed to calculate $P_\theta(\omega)$ is given by

$$\begin{aligned}\overline{i(t)i(t+\tau)} - \left(\overline{i(t)}\right)^2 &= (Ge)^2|\hat{\mathcal{E}}_{lo}|^2\xi^2 4\langle : \Delta\hat{A}_\theta(t)\Delta\hat{A}_\theta(t+\tau) : \rangle \\ &\quad + (Ge)^2|\hat{\mathcal{E}}_{lo}|^2\delta(\tau).\end{aligned} \tag{6.41}$$

Then, after taking the Fourier transform (6.40), we have

$$P_\theta(\omega) = P^\theta_{hom}(\omega) + P_{shot}, \tag{6.42}$$

where

$$P^\theta_{hom}(\omega) = (Ge)^2|\hat{\mathcal{E}}_{lo}|^2\xi^2 4\int_0^\infty d\tau \cos\omega\tau \lim_{t\to\infty}\langle : \Delta\hat{A}_\theta(t)\Delta\hat{A}_\theta(t+\tau) : \rangle, \tag{6.43a}$$

$$P_{shot}(\omega) = (Ge)^2|\hat{\mathcal{E}}_{lo}|^2/2\pi, \tag{6.43b}$$

and the *ideal source-field spectrum of squeezing* is defined by

$$\begin{aligned} S_\theta(\omega) &= \frac{P_\theta(\omega) - P_{shot}}{P_{shot}} \\ &= (2\kappa)8\int_0^\infty d\tau \cos\omega\tau \lim_{t\to\infty} \langle : \Delta\hat{A}_\theta(t)\Delta\hat{A}_\theta(t+\tau) : \rangle, \end{aligned} \tag{6.44}$$

with ξ replaced by $\sqrt{2\kappa}$, which is the scaling required to convert $\Delta\tilde{a}(t)$ [in equation (6.35)] into units of total photon flux out of the cavity. The spectrum of photocurrent fluctuations is given in terms of the spectrum of squeezing by

$$P_\theta(\omega)/P_{shot} = 1 + S_\theta(\omega). \tag{6.45}$$

6.3 Vacuum fluctuations

Our definition of the spectrum of squeezing has been based on a description of homodyne detection. From this point of view we have a clear idea of what the spectrum of squeezing means in terms of photocurrent fluctuations; since the photocurrent is a classical stochastic quantity we can conjure up a mental picture of its fluctuations. It is tempting to extend this picture to the field, and regard the fluctuating current to be a direct "mapping" (measurement) of fluctuations in the quantized field. Here we must exercise some caution. Certainly we can visualize the field if it carries classical (for example, thermal) fluctuations; but, can we construct a mental picture of nonclassical fluctuations in the field to match our picture of photocurrent fluctuations? If we can, what is the basis for this picture; what is the mathematical correspondence between the fluctuations in the photocurrent and the fluctuations in the field?

To discuss these questions we must amend our definition of the spectrum of squeezing a little. Note that the field entering (6.44) is the source field $\Delta\hat{A}_\theta(t)$ alone; the free field [Eq. (1.54)] has been omitted. This is why we used the qualification *source field* spectrum of squeezing. The omission does not matter if the free field is in the vacuum state because of the normal ordering and time ordering specified in (6.44). But the free field must be present for the discussion that follows. We therefore consider the *ideal spectrum of squeezing* given by

$$\begin{aligned} S_\theta(\omega) = 8\int_0^\infty d\tau \cos\omega\tau \lim_{t\to\infty} \Big\langle : \Big[\sqrt{c/2L'}\,\hat{F}_\theta(t) + \sqrt{2\kappa}\,\Delta\hat{A}_\theta(t)\Big] \\ \times \Big[\sqrt{c/2L'}\,\hat{F}_\theta(t+\tau) + \sqrt{2\kappa}\,\Delta\hat{A}_\theta(t+\tau)\Big] : \Big\rangle, \end{aligned} \tag{6.46}$$

where

$$\hat{F}_\theta = \tfrac{1}{2}(\hat{\tilde{f}} e^{-i\theta} + \hat{\tilde{f}}^\dagger e^{i\theta}), \tag{6.47}$$

with

$$\hat{\tilde{f}} = (\sqrt{\gamma_{a1}}\,\tilde{r}_{a1f} + \sqrt{\gamma_{a2}}\,\tilde{r}_{a2f} + \sqrt{\gamma_{a\alpha}}\,\tilde{r}_{a\alpha f})/\sqrt{2\kappa}; \tag{6.48a}$$

alternatively,

$$\hat{\tilde{f}} = \sum_\omega \sqrt{\frac{\omega_C + \omega}{\omega_C}}\,\hat{f}_\omega e^{-i\omega\tau}. \tag{6.48b}$$

The operators $\hat{\tilde{f}}$ are reservoir mode operators like those appearing in the expansion (1.67b). In (6.48a) there is a sum of three pieces because we allow for three sources of loss from the oscillator cavity: from either of two partially transmitting mirrors (γ_{a1} and γ_{a2}), and by absorption in the crystal (γ_α); the free field in (6.46) is a composite field that accounts for all output channels. If this is confusing, just set $\gamma_{a2} = \gamma_\alpha = 0$, $\gamma_{a1} = 2\kappa$ to obtain results for a cavity with one output mirror.

The ideal source-field spectrum of squeezing (6.44) is the quantity computed in the work of Walls and coworkers [6.7–6.9]. It is equivalently given by (6.46) when the free field is in the vacuum state. We are going to relate (6.46) to the expression without normal ordering and time ordering widely used in the work of others.

The question we posed above can be answered in the affirmative – we can construct a mental picture of the fluctuations in the quantized field. We do this using the quantum-classical correspondence (Lecture 4). Equation (6.46) states that the spectrum of squeezing is the Fourier transform of the normal-ordered, time-ordered correlation function for quadrature phase operators of the quantized field $\sqrt{c/2L'}\,\hat{\tilde{f}} + \sqrt{2\kappa}\,\Delta\tilde{a}$. Since it is the P representation that evaluates normal-ordered, time-ordered correlation functions as "classical" integrals, the Fokker-Planck equation in the P representation (and its associated stochastic differential equation) provides the desired visualization of the fluctuating field. But there is a problem. We have stated that the P distribution for a squeezed state does not exist as a well-behaved function. This fact is revealed in the Fokker-Planck equation [Eq. (4.72b) with $\sigma = 1$] which does not have positive semidefinite diffusion. It seems, then, that the P representation cannot be used to construct a classical picture of the field. The solution is to use either the Q or the Wigner representation, since in these representations the Fokker-Planck equations do have positive semidefinite diffusion [$\sigma = -1$ and $\sigma = 0$ in (4.72b)]. However, if we do this we must change the operator ordering in the expression for the spectrum of squeezing. Let us reorder the operators to clarify the connection between the spectrum of squeezing and the Wigner stochastic representation of the fluctuating field.

The normal-ordered, time-ordered averages that appear in the expression for the spectrum of squeezing are related to averages without normal ordering and time ordering by

$$\langle :\hat{F}_\theta(t)\hat{F}_\theta(t+\tau): \rangle = \langle \hat{F}_\theta(t)\hat{F}_\theta(t+\tau)\rangle + \tfrac{1}{4}\langle [\hat{\tilde{f}}(t+\tau)e^{-2i\theta} + \hat{\tilde{f}}^\dagger(t+\tau), \hat{\tilde{f}}(t)]\rangle, \tag{6.49a}$$

$$\langle :\hat{F}_\theta(t)\Delta\hat{A}_\theta(t+\tau): \rangle = \langle \hat{F}_\theta(t)\Delta\hat{A}_\theta(t+\tau)\rangle + \tfrac{1}{4}\langle [\Delta\tilde{a}(t+\tau)e^{-2i\theta} + \Delta\tilde{a}^\dagger(t+\tau), \hat{\tilde{f}}(t)]\rangle, \tag{6.49b}$$

$$\langle :\Delta\hat{A}_\theta(t)\hat{F}_\theta(t+\tau): \rangle = \langle \Delta\hat{A}_\theta(t)\hat{F}_\theta(t+\tau)\rangle + \tfrac{1}{4}\langle [\hat{\tilde{f}}(t+\tau)e^{-2i\theta} + \hat{\tilde{f}}^\dagger(t+\tau), \Delta\tilde{a}(t)]\rangle, \tag{6.49c}$$

$$\langle :\Delta\hat{A}_\theta(t)\Delta\hat{A}_\theta(t+\tau): \rangle = \langle \Delta\hat{A}_\theta(t)\Delta\hat{A}_\theta(t+\tau)\rangle + \tfrac{1}{4}\langle [\Delta\tilde{a}(t+\tau)e^{-2i\theta} + \Delta\tilde{a}^\dagger(t+\tau), \Delta\tilde{a}(t)]\rangle. \tag{6.49d}$$

The free-field commutator that appears on the right-hand side of (6.49a) can be evaluated using (6.48b) and the boson commutation relations for the $\hat{f}_\omega$:

$$\begin{aligned}\langle [\hat{\tilde{f}}(t+\tau) + \hat{\tilde{f}}^\dagger(t+\tau), \hat{\tilde{f}}(t)]\rangle &= \sum_\omega \frac{\omega_C+\omega}{\omega_C}\langle [f_\omega^\dagger, f_\omega]\rangle e^{i\omega\tau} \\ &= -(L'/\pi c)\int_{-\infty}^{\infty} d\omega \frac{\omega_C+\omega}{\omega_C} e^{i\omega\tau} \\ &= -(2L'/c)\delta(\tau),\end{aligned} \tag{6.50}$$

where we have used the quasimonochromatic condition $\omega \ll \omega_C$. Now we express the expectations of commutators between source-field and free-field operators in terms of source-field operators alone using (6.48a) and the correlation functions (1.75):

$$\begin{aligned}&\langle [\Delta\tilde{a}(t+\tau)e^{-2i\theta} + \Delta\tilde{a}^\dagger(t+\tau), \hat{\tilde{f}}(t)]\rangle \\ &= \begin{cases} -\sqrt{2\kappa}\sqrt{2L'/c}\Big[\langle [\Delta\tilde{a}(t+\tau)e^{-2i\theta} + \Delta\tilde{a}^\dagger(t+\tau), \Delta\tilde{a}(t)]\rangle\Big] & \tau > 0, \\ -\tfrac{1}{2}\sqrt{2\kappa}\sqrt{2L'/c}\Big[\langle [\Delta\tilde{a}(t+\tau)e^{-2i\theta} + \Delta\tilde{a}^\dagger(t+\tau), \Delta\tilde{a}(t)]\rangle\Big] & \tau = 0, \end{cases}\end{aligned} \tag{6.51a}$$

$$\begin{aligned}&\langle [\hat{\tilde{f}}(t+\tau)e^{-2i\theta} + \hat{\tilde{f}}^\dagger(t+\tau), \Delta\tilde{a}(t)]\rangle \\ &= \begin{cases} 0 & \tau > 0, \\ -\tfrac{1}{2}\sqrt{2\kappa}\sqrt{2L'/c}\Big[\langle [\Delta\tilde{a}(t+\tau)e^{-2i\theta} + \Delta\tilde{a}^\dagger(t+\tau), \Delta\tilde{a}(t)]\rangle\Big] & \tau = 0. \end{cases}\end{aligned} \tag{6.51b}$$

Notice that the commutator appearing on the right-hand sides in (6.51a) and (6.51b) is the same, with opposite sign, as the commutator on the right-hand side of (6.49d). Thus, when we substitute (6.49a)–(6.49d) into (6.46), and use (6.50), and (6.51a) and (6.51b), we find

$$\begin{aligned} \mathcal{S}_\theta(\omega)+1 = 8\int_0^\infty d\tau \cos\omega\tau \lim_{t\to\infty}\Big\langle\Big[\sqrt{c/2L'}\,\hat{F}_\theta(t)+\sqrt{2\kappa}\,\Delta\hat{A}_\theta(t)\Big] \\ \times\Big[\sqrt{c/2L'}\,\hat{F}_\theta(t+\tau)+\sqrt{2\kappa}\,\Delta\hat{A}_\theta(t+\tau)\Big]\Big\rangle. \end{aligned} \tag{6.52}$$

The $+1$ in $\mathcal{S}_\theta(\omega)+1$ comes from the Fourier transform of the δ-function (6.50), which originates in the correlation function $\langle\hat{\hat{f}}(t)\hat{\hat{f}}^\dagger(t+\tau)\rangle$. This δ-function represents a contribution from the vacuum fluctuations in the free field (6.48). When we compare (6.52) with the photocurrent spectrum $P_\theta(\omega)$ [Eq. (6.45)] we see that the contribution from the vacuum fluctuations corresponds to the shot noise component (normalized to unity) of the photocurrent fluctuations. Our derivation of $P_\theta(\omega)$ showed that shot noise arises from the self-correlation of the individual pulses that make up the photocurrent. With $P_\theta(\omega)/P_{shot} = \mathcal{S}_\theta(\omega)+1$ calculated from (6.52), we are now permitted an interpretation in which the shot noise is associated with vacuum fluctuations in the reservoir fields. This interpretation is made clearer by rewriting (6.52) in the form

$$\begin{aligned} \mathcal{S}_\theta(\omega)+1 = 4(\pi c/L')\frac{1}{2\pi}\int_{-\infty}^\infty d\tau e^{i\omega\tau} \lim_{t\to\infty}\Big\langle\Big[\hat{F}_\theta(t)+\sqrt{2\kappa}\sqrt{2L'/c}\,\Delta\hat{A}_\theta(t)\Big] \\ \times\Big[\hat{F}_\theta(t+\tau)+\sqrt{2\kappa}\sqrt{2L'/c}\,\Delta\hat{A}_\theta(t+\tau)\Big]\Big\rangle, \end{aligned} \tag{6.53}$$

where we have used

$$\Big\langle\Big[\hat{F}_\theta(t)+\sqrt{2\kappa}\sqrt{2L'/c}\,\Delta\hat{A}_\theta(t),\hat{F}_\theta(t+\tau)+\sqrt{2\kappa}\sqrt{2L'/c}\,\Delta\hat{A}_\theta(t+\tau)\Big]\Big\rangle = 0. \tag{6.54}$$

The integral is now a Fourier transform.

The averages that appear on the right-hand side of (6.53) can be calculated as phase-space averages in the Wigner representation. Actually, the Wigner representation gives correlation functions in symmetrized time order; therefore, strictly, the spectra of quadrature phase amplitude fluctuations calculated in the Wigner representation would give the average of (6.53) and the same expression with the operator order reversed. But, from (6.54), the time order is unimportant. Thus, we can write

$$\begin{aligned} &\mathcal{S}_\theta(\omega)+1 \\ &= 4\times(\pi c/L')\times\left(\begin{array}{c}\text{variance of quadrature phase amplitude}\\ \text{fluctuations per unit bandwidth (in photon number units)}\\ \text{in the Wigner stochastic representation of the field}\end{array}\right). \end{aligned} \tag{6.55}$$

The factor of 4 scales the quadrature variance of 1/4 per mode to unity, and $\pi c/L'$ is the mode spacing in frequency space.

The spectra of Wigner stochastic fluctuations computed for the intracavity field and the cavity output field are quite different. This is because these spectra include a contribution associated with vacuum fluctuations. The cavity acts as a filter which suppresses vacuum fluctuations at frequencies outside the cavity linewidth. Thus, spectra computed for the intracavity field combine this suppression with squeezing induced effects. If the visualization of field fluctuations is built around normal-ordered, time-ordered correlation functions this difference between spectra inside and outside the cavity only arises when the free fields carry a real photon flux (when the reservoirs are not in the vacuum state).

6.4 Squeezing spectra for the degenerate parametric oscillator

We now put a number of the tools we have learned together to calculate something useful and nontrivial. The spectrum of squeezing given by (6.44) characterizes the photocurrent fluctuations in homodyne detection of a source field $\sqrt{2\kappa}\,a(t)$. To calculate this spectrum we need the correlation function that appears in the integrand on the right-hand side. The correlation function may be calculated using one of the methods of analysis discussed in Lectures 3 and 4. We will calculate the spectrum of squeezing for the output from a degenerate parametric oscillator modeled by the source master equation (2.63). We assume the oscillator cavity has only one output mirror and there are no losses in the nonlinear crystal. Under these conditions (6.44) is not just the *ideal* spectrum of squeezing, but the spectrum actually measured by a detector monitoring the cavity output. The correlation function we need can be obtained from the Fokker-Planck equations (4.72) that describe the subharmonic mode fluctuations below threshold.

The drift terms (first derivatives) in (4.72a) and (4.72b) correspond to the deterministic equations

$$\dot{\tilde{z}}_1 = -\kappa(1-\lambda)\tilde{z}_1, \qquad \dot{\tilde{z}}_2 = -\kappa(1+\lambda)\tilde{z}_2; \tag{6.56}$$

the terms $+\kappa\lambda\tilde{z}_1$ and $-\kappa\lambda\tilde{z}_2$ describe the amplification and deamplification of quadrature phase amplitudes seen in the parametric amplifier results (6.16) and (6.22). Below threshold the gain for $\tilde{z}_1$ is less than the loss; therefore, below threshold the fluctuations do not initiate the growth of a mean field amplitude. The fluctuations, however, experience a phase-dependent decay, which leads to a phase-dependence in the mean deviation of the fluctuations from steady state. Before we calculate the spectrum of squeezing let us simply calculate the variance of the field fluctuations. We consider fluctuations in the quadrature phase operators $\hat{X} \equiv \hat{A}_\theta$, $\hat{Y} \equiv \hat{A}_{\theta+\pi/2}$, with $\theta = \psi/2$, where ψ is the phase appearing in the scaling relation (4.55). Since $\langle\tilde{a}\rangle = \langle\tilde{a}^\dagger\rangle = 0$, we may use (6.20) and the scaling relations (4.55)–(4.57) to write

$$\begin{aligned}(\Delta X)^2 &= \tfrac{1}{4}\langle(\tilde{a}e^{-i\psi} + \tilde{a}^\dagger e^{i\psi})^2\rangle \\ &= (2/\xi)\Big[(\overline{\tilde{z}_1\tilde{z}_1})_{\tilde{X}_\sigma} + \tfrac{1}{4}\sigma\Big],\end{aligned} \tag{6.57a}$$

where we use the fact that phase-space averages in the P, Wigner, and Q representations give normal-ordered, symmetric-ordered, and antinormal-ordered operator averages, respectively [recall that σ distinguishes between the representations – Eq. (4.54)]; the subscript indicates the distribution used in the calculation of the average. A similar calculation gives

$$(\Delta Y)^2 = (2/\xi)\Big[(\overline{\tilde{z}_2\tilde{z}_2})_{\tilde{Y}_\sigma} + \tfrac{1}{4}\sigma\Big]. \tag{6.57b}$$

The variances of the Gaussian steady-state solutions to (4.72a) and (4.72b) give [compare the Fokker-Planck equation (4.32) and its solution (4.33)]

$$(\Delta X)_{ss} = \frac{1}{2}\sqrt{\frac{1}{1-\lambda}}, \tag{6.58a}$$

$$(\Delta Y)_{ss} = \frac{1}{2}\sqrt{\frac{1}{1+\lambda}}. \tag{6.58b}$$

We see that fluctuations in the deamplified Y quadrature phase amplitude are less than those in the vacuum state [Eq. (6.25)]; at threshold $\Delta Y = \frac{1}{2}(1/\sqrt{2}) < \frac{1}{2}$. Fluctuations in the amplified X quadrature phase amplitude diverge as threshold is approached. [The divergence signals the breakdown of the system size expansion].

Although the choice of representation enters explicitly into (6.57a) and (6.57b), the results for (ΔX) and (ΔY) are independent of σ. This, of course, is as it should be; different representations cannot produce different answers for the same operator average. The term $\frac{1}{4}\sigma$ in (6.57a) and (6.57b) cancels the σ dependence in the variances for the phase-space variables. Since $\sigma = 0$ corresponds to the Wigner representation, we see why it is the contours of this representation that relate most directly to pictures like those in Fig. 6.2.

Now, what is the quantum state of the subharmonic field? From (6.58) it is clear that it is not a minimum uncertainty state and therefore it is not a squeezed state – $(\Delta X)_{ss}(\Delta Y)_{ss} = \frac{1}{4}\big(\sqrt{1-\lambda^2}\big)^{-1} > \frac{1}{4}$. When Milburn and Walls [6.14] first analyzed this model, the minimum value of ΔY predicted by (6.58b) – $\Delta Y = \frac{1}{2}(1/\sqrt{2})$ for $\lambda = 1$ – was something of a disappointment; ΔY is only reduced by the factor $1/\sqrt{2}$ from its value in the vacuum state. But, fortunately, this pessimistic outlook is the result of an oversimplified analysis. It is important to recognize that the cavity mode that carries the subharmonic field is really a quasimode (it has a linewidth); also, that it is photocurrent fluctuations that are actually observed, and these may not correspond to the variances calculated above. Indeed from our analysis of homodyne detection we see that in place of (6.58a) and (6.58b) we must look at fluctuation amplitudes $\frac{1}{2}\sqrt{1+S(\omega,\theta)}$, for $\theta=\psi/2$ and $\theta=\psi/2+\pi/2$ [Eq. (6.45)], where $S(\omega,\theta)$ is the spectrum of squeezing. Setting $\theta=\psi/2$ in (6.44), we have

$$\begin{aligned}S_X(\omega) &= (2\kappa)8\int_0^\infty d\tau\cos\omega\tau\lim_{t\to\infty}\Big\langle :\tfrac{1}{2}\big[\tilde{a}(t)e^{-i\psi/2}+\tilde{a}^\dagger(t)e^{i\psi/2}\big]\\ &\quad\times\tfrac{1}{2}\big[\tilde{a}(t+\tau)e^{-i\psi/2}+\tilde{a}^\dagger(t+\tau)e^{i\psi/2}\big]:\Big\rangle\\ &= (2\kappa)8\int_0^\infty d\tau\cos\omega\tau(2/\xi)\lim_{t\to\infty}\big(\overline{\tilde{z}_1(t)\tilde{z}_1(t+\tau)}\big)_{\bar{\tilde{X}}_{+1}},\end{aligned} \tag{6.59a}$$

where $S_X(\omega)\equiv S(\psi/2,\omega)$; in a similar manner, with $\theta=\psi/2+\pi/2$,

$$S_Y(\omega) = (2\kappa)8\int_0^\infty d\tau\cos\omega\tau(2/\xi)\lim_{t\to\infty}\big(\overline{\tilde{z}_2(t)\tilde{z}_2(t+\tau)}\big)_{\bar{\tilde{Y}}_{+1}}, \tag{6.59b}$$

where $S_Y(\omega)\equiv S(\psi/2+\pi/2,\omega)$. The P representation ($\sigma=+1$) is used in these expressions to compute the phase-space correlation functions that give the normal-ordered, time-ordered averages required by (6.44) (strictly, the ***positive P representation*** is needed to make sense of the negative diffusion [6.15]). The correlation functions calculated in the P representation are

$$\lim_{t\to\infty}\big(\overline{\tilde{z}_1(t)\tilde{z}_1(t+\tau)}\big)_{\bar{\tilde{X}}_{+1}} = (\xi/2)\frac{1}{4}\frac{\lambda}{1-\lambda}e^{-\kappa(1-\lambda)|\tau|}, \tag{6.60a}$$

$$\lim_{t\to\infty}\big(\overline{\tilde{z}_2(t)\tilde{z}_2(t+\tau)}\big)_{\bar{\tilde{Y}}_{+1}} = -(\xi/2)\frac{1}{4}\frac{\lambda}{1+\lambda}e^{-\kappa(1+\lambda)|\tau|}, \tag{6.60b}$$

and hence,

$$S_X(\omega) = \frac{\lambda}{1-\lambda}\frac{4\kappa^2(1-\lambda)}{[\kappa(1-\lambda)]^2+\omega^2}, \tag{6.61a}$$

$$S_Y(\omega) = -\frac{\lambda}{1+\lambda}\frac{4\kappa^2(1+\lambda)}{[\kappa(1+\lambda)]^2+\omega^2}. \tag{6.61b}$$

Thus, the fluctuation amplitudes defined via the spectrum of squeezing are

$$\tfrac{1}{2}\sqrt{1+S_X(\omega)} = \frac{1}{2}\sqrt{\frac{[\kappa(1+\lambda)]^2+\omega^2}{[\kappa(1-\lambda)]^2+\omega^2}}, \tag{6.62a}$$

$$\tfrac{1}{2}\sqrt{1+S_Y(\omega)} = \frac{1}{2}\sqrt{\frac{[\kappa(1-\lambda)]^2+\omega^2}{[\kappa(1+\lambda)]^2+\omega^2}}. \tag{6.62b}$$

We now find a close connection with minimum uncertainty squeezed states. The minimum uncertainty condition $\frac{1}{2}\sqrt{1+S_X(\omega)}\frac{1}{2}\sqrt{1+S_Y(\omega)}=\frac{1}{4}$ is satisfied at each frequency. Furthermore, at line center the squeezing becomes perfect as threshold is approached, with $\frac{1}{2}\sqrt{1+S_Y^{\lessgtr}(0)}\to 0$ and $\frac{1}{2}\sqrt{1+S_X^{\lessgtr}(0)}\to\infty$ as $\lambda\to 1$.

6.5 Photoelectron counting for the degenerate parametric oscillator

We should say something about the direct photoelectron counting distribution for the squeezed output of the degenerate parametric oscillator. Homodyne detection is used to observe the phase-sensitive amplitude fluctuations of squeezed light. Of course, it is also possible to omit the local oscillator and count the photoelectrons produced by the squeezed light alone. We will not spend time on algebraic details – as we already stated for the case of resonance fluorescence (Fig. 5.3), these are fairly involved. The physics contained in the results is quite transparent without going into the mathematics used in their derivation.

Photoelectron counting distributions for the degenerate parametric oscillator have been calculated by Vyas and Singh [6.16], and Vyas and DeBrito [6.17] using an analytical method based on the positive P representation, and by Wolinsky and Carmichael [6.18] using an numerical approach based on the decomposition of master equation dynamics that we will be discussing in the remaining four lectures. Two examples from the work of Wolinsky and Carmichael appear in Fig. 6.4. The counting distribution for operation well below threshold [Fig. 6.4(a)] shows only even numbers of photoelectron counts. The even counts result because the subharmonic photons are produced in pairs inside the cavity. Well below threshold the pairs are created at a slow rate compared with the rate $(2\kappa)^{-1}$ at which photons leave the cavity. Thus, photons emerge from the cavity in distinct pairs, with the two photons of each pair separated, on average, by a time $(2\kappa)^{-1}$. The photoelectron counting distribution reflects this fact if the counting time is sufficiently long that it is very unlikely that the turn-on and turn-off of the counting interval will split a pair. This is the regime of spontaneous parametric down conversion. Close to threshold stimulated events become more important [Fig. 6.4(b)]. Pairs are created inside the cavity at a rate comparable to $(2\kappa)^{-1}$. Under these conditions photons do not leave the cavity as distinct pairs. The average time $(2\kappa)^{-1}$ separating the members of a pair as they leave the cavity is similar to the average time separating successive pair creations. The number of photoelectrons counted in a fixed interval can then be even or odd; it becomes quite likely that the turn-on and turn-off of the counting interval will split a pair.

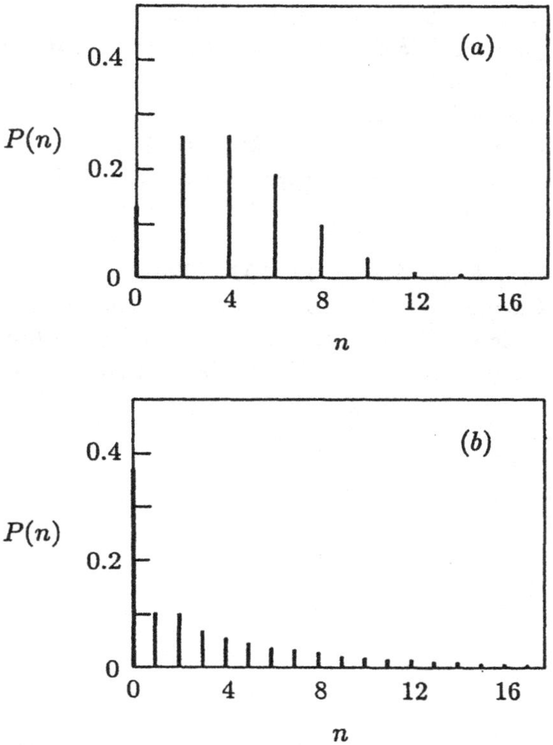

Fig. 6.4. Photoelectron counting distributions for the degenerate parametric oscillator operated below threshold. (a) 90% below threshold ($\lambda = 0.1$), (b) 10% below threshold ($\lambda = 0.9$). The detector has unit quantum efficiency. In (a) the counting time is $T = 200 \times (2\kappa)^{-1}$, and in (b), $0.5 \times (2\kappa)^{-1}$.

References

[6.1] D. F. Walls, *Nature* **306**, 141 (1983).

[6.2] H. P. Yuen, *Phys. Rev.* A **13**, 2226 (1976).

[6.3] *Journal of Modern Optics*, Vol. **34**, Nos. 6/7 (1987).

[6.4] *Journal of the Optical Society of America* B, Vol. **4** (1987).

[6.5] L.-A. Wu, M. Xiao, and H. J. Kimble, *J. Opt. Soc. Am.* B **4**, 1465 (1987).

[6.6] H. J. Carmichael, *J. Opt. Soc. Am.* B **4**, 1588 (1987).

[6.7] M. J. Collett, D. F. Walls, and P. Zoller, *Optics Commun.* **52**, 145 (1984).

[6.8] M. J. Collett and D. F. Walls, *Phys. Rev.* A **32**, 2887 (1985).

[6.9] M. D. Reid and D. F. Walls, *Phys. Rev.* A **32**, 396 (1985); *Phys. Rev.* A **34**, 4929 (1986).

[6.10] C. Caves and B. L. Schumaker, *Phys. Rev.* A **31**, 3068 (1985); *ibid*, 3093 (1985).

[6.11] H. P. Yuen and V. W. S. Chan, *Opt. Lett.* **8**, 177 (1983).
[6.12] H. P. Yuen and J. H. Shapiro, *IEEE Trans. Inf. Theory* **IT-26**, 78 (1980).
[6.13] B. Yurke, *Phys. Rev.* A **29**, 408 (1984).
[6.14] G. Milburn and D. F. Walls, *Optics Commun.* **39**, 401 (1981).
[6.15] P. D. Drummond and C. W. Gardiner, *J. Phys.* A **13**, 2353 (1980).
[6.16] R. Vyas and S. Singh, *Opt. Lett.* **14**, 1110 (1989); *Phys. Rev.* A **40**, 5147 (1989).
[6.17] R. Vyas and A. L. DeBrito, *Phys. Rev.* A **42**, 592 (1990).
[6.18] M. Wolinsky and H. J. Carmichael, "Photoelectron Counting Statistics for the Degenerate Parametric Oscillator," in *Coherence and Quantum Optics VI*, ed. by J. H. Eberly, L. Mandel, and E. Wolf, Plenum: New York, 1989, pp. 1239ff.

Lecture 7 – Quantum Trajectories I

We are now able to begin the enterprise towards which the previous six lectures have been heading. We are going to develop a new way of thinking about and analyzing the master equation for a photoemissive source. The character of the new approach can be appreciated by considering an analogy with classical statistical physics. In classical statistical physics there are two ways of approaching the dynamical evolution of a system. In the first the system is described by a probability distribution and a Fokker-Planck equation, or its equivalent, generates the evolution in time. In the second the system is describe by an ensemble of noisy trajectories and a set of stochastic differential equations is used to generate the trajectories. The quantum-classical correspondence (Sects. 4.1–4.3) allows both of these methods to be used to analyze a source master equation. But the usefulness of this method is limited. It is limited ultimately by the fact that at a fundamental level, quantum dynamics does not fit the classical statistics mold. It is actually rare that an operator master equation is converted into a Fokker-Planck equation under the quantum-classical correspondence. Most often the system size expansion (small quantum noise assumption) is used to "shoehorn" the quantum dynamics into a classical form. If this cannot be done, then we always have the operator master equation itself, which might be solved directly, using a computer if necessary. The master equation is an equation for the density operator – the quantum mechanical version of a probability distribution. What we do not seem to have is a quantum mechanical version of the stochastic trajectories. Certainly, we can obtain operator stochastic differential equations from the Heisenberg equations of motion [7.1]. But what about the *pictures* that classical stochastic trajectories evoke? Can we build a formalism that produces similar pictures, pictures of quantum stochastic trajectories? We are going to see how this can be done. The mathematics we will use is essentially that developed by Davies in his theory of continuous quantum measurement [7.2]. The connections between the next four lectures and this theory of quantum measurement are very close. However, our perspective is different from that taken by Davies' theory. We focus our attention on the quantum dynamics of the photoemissive source, not on the interaction between its radiated field and some detector. We will certainly be instructed by the theory of photoelectric detection. But because the source is itself an open system, we may regard the detector as a device monitoring (and in a sense selecting) what the source does, and not interfering with the source dynamics in any direct way. The difference is clear when it is viewed against the claim that Davies' theory corrects deficiencies in the standard theory of photoelectric detection [7.3–7.5]. We, in fact, use only the standard theory of photoelectric detection. For a photoemissive source the standard theory contains the Davies' mathematical language buried within itself. We must simply extricate it and then put it to use.

7.1 Exclusive and nonexclusive photoelectron counting probabilities

In Lecture 5 we met the photoelectron counting distribution for a quantized field in the form [Eqs. (5.39)]

$$p(n,t,T) = \left\langle : \frac{\left[\xi \int_t^{t+T} dt' \hat{E}^{(-)}(t')\hat{E}^{(+)}(t')\right]^n}{n!} \times \exp\left[-\xi \int_t^{t+T} dt' \hat{E}^{(-)}(t')\hat{E}^{(+)}(t')\right] : \right\rangle. \tag{7.1}$$

The constant ξ is the product of the detector quantum efficiency and a factor that converts the electric field intensity into a photon flux, and the notation $\langle : : \rangle$ indicates that all operators are to be written in normal order and time order. After expanding the exponential, (7.1) expresses $p(n,t,T)$ as a rather complicated series of integrals over the probability densities

$$w_m(t_1,t_2,\ldots,t_m) = \xi^m \langle \hat{E}^{(-)}(t_1)\cdots\hat{E}^{(-)}(t_m)\hat{E}^{(+)}(t_m)\cdots\hat{E}^{(+)}(t_1)\rangle. \tag{7.2}$$

These are called the *nonexclusive probability densities*, or the multicoincidence rates, for photoelectric counting [7.6]. The nonexclusive probability $w_m(t_1,t_2,\ldots,t_m)\Delta t_1\Delta t_2\ldots\Delta t_m$ is the probability that one photoelectron is emitted in each of the nonoverlapping intervals $[t_1,t_1+\Delta t_1), [t_2,t_2+\Delta t_2),\ldots,[t_m,t_m+\Delta t_m)$, where $t_1 < t_2 < \cdots < t_m$. The nonexclusive probability densities are proportional to the multi-time correlation functions introduced by Glauber in his quantum theory of optical coherence [7.7, 7.8]. The special significance of the Glauber-Sudarshan P representation mentioned in Sect. 4.1 is tied to its usefulness in calculating these nonexclusive probability densities or multi-time coincidence rates.

The *nonexclusive* character of the probability densities (7.2) comes from the fact that they place no conditions on what might happen at times in between the m infinitesimal intervals during which the specified photoelectron emissions occur. For example, perhaps there are ten photoelectrons emitted in the interval $[t_1+\Delta t_1, t_2)$, perhaps there are five, or perhaps none; nonexclusive probabilities make no distinction between these possibilities.

There is only the statement that m photoelectron emissions occur at m specified times; all sequences involving various combinations of additional emissions in between these times are summed together in the definition of the nonexclusive probabilities.

The nonexclusive probability densities are rather straightforward to measure since they are just multi-time coincidence rates. It is only necessary to gate a detector "on" and "off" to define the m active intervals, and then record when a photoelectric pulse is observed on all m occasions. On the other hand, the photoelectron counting distribution has a very complicated form when expressed in terms of nonexclusive probability densities.

There is a much simpler expression for the photoelectron counting distribution in terms of exclusive probability densities:

$$p(n,t,t+T) = \int_t^{t+T} dt_n \int_t^{t_n} dt_{n-1} \cdots \int_t^{t_2} dt_1\, \wp_n(t_1,t_2,\ldots,t_n;[t,t+T]). \tag{7.3}$$

The $\wp_n(t_1,t_2,\ldots,t_n;[t,t+T])$ are the *exclusive probability densities* for photoelectron counting; $\wp_n(t_1,t_2,\ldots,t_n;[t,t+T])\Delta t_1\ \Delta t_2 \ldots \Delta t_n$ is the probability that n photoelectrons are emitted in the observation interval $[t,t+T]$, one in each of the nonoverlapping intervals $[t_1,t_1+\Delta t_1), [t_2,t_2+\Delta t_2),\ldots,[t_n,t_n+\Delta t_n)$, where $t_1 < t_2 < \cdots < t_n$. In this definition those events in which emissions occur in between the n specified intervals are excluded – giving an *exclusive* probability. The stochastic process that describes the photoelectron emission sequences is completely defined, either by the full hierarchy of nonexclusive probability densities, or by the full hierarchy of exclusive probability densities. Traditionally quantum optics has drawn most of its conceptual framework, and also its calculational methods, from a consideration of the nonexclusive probability densities. We are now going to shift our attention to the exclusive probability densities.

From the definition of the exclusive probability densities the expression (7.3) for the photoelectron counting distribution has a rather obvious interpretation. The integrals on the right-hand side are simply summing up over all the possible sequences of n photoelectron emissions that can occur in the interval $[t,t+T]$. The only question that remains is how do we calculate the exclusive probability densities. The answer is not obvious because to confound the simple form of (7.2), we now have a complicated relationship between the exclusive probability densities and the nonexclusive probability densities [7.9]:

$$\begin{aligned}&\wp_m(t_1,t_2,\ldots,t_m;[t,t+T])\\ &\quad= \sum_{r=0}^{\infty} \frac{(-1)^r}{r!} \int_t^{t+T} dt'_r \int_t^{t+T} dt'_{r-1} \ldots \int_t^{t+T} dt'_1 \langle : \hat{I}(t'_r)\hat{I}(t'_{r-1}) \ldots \\ &\qquad \ldots \hat{I}(t'_1)\hat{I}(t_m)\ldots\hat{I}(t_2)\hat{I}(t_1) : \rangle\end{aligned}$$

$$= \Big\langle : e^{-\hat{\Omega}(t,t+T)} \hat{I}(t_m) \dots \hat{I}(t_2)\hat{I}(t_1) : \Big\rangle$$
$$= \Big\langle : e^{-\hat{\Omega}(t+T,t_m)} \hat{I}(t_m) \dots \hat{I}(t_2) e^{-\hat{\Omega}(t_2,t_1)} \hat{I}(t_1) e^{-\hat{\Omega}(t_1,t)} : \Big\rangle, \tag{7.4}$$

where

$$\hat{\Omega}(t_i, t_j) = \int_{t_j}^{t_i} dt' \hat{I}(t'), \tag{7.5}$$

and $\hat{I}(t)$ is the photon flux operator

$$\hat{I}(t) = \xi \hat{E}^{(-)}(t)\hat{E}^{(+)}(t). \tag{7.6}$$

Before we look at the way in which this quantity might be evaluated, let us say a little more about the difference between nonexclusive and exclusive probability densities.

7.2 The distribution of waiting times

In Lecture 5 we discussed two related quantities: the (normalized) second-order correlation function $g^{(2)}(\tau)$ and the waiting-time distribution $w(\tau)$. These quantities both dealt with probabilities for observing two photoelectron emissions separated by a time delay τ. But there was a difference; the definition of $w(\tau)$ required that there be no additional emissions during the interval τ [Eq. (5.31)], while $g^{(2)}(\tau)$ was defined in terms of a coincidence counting probability that does not make this qualification. Here we have the simplest example of the distinction between nonexclusive and exclusive probabilities. To be more precise, in the present notation the normalized second-order correlation function is defined by

$$g^{(2)}(t, t+\tau) = w_2(t, t+\tau)/[w_1(t)w_1(t+\tau)]; \tag{7.7}$$

it is a normalized version of the nonexclusive probability density $w_2(t, t+\tau)$. The waiting-time distribution is defined in terms of the conditional exclusive probability densities

$$\wp_m(t_1, t_2, \dots, t_m | t_0) = \wp_{m+1}\big(t_0, t_1, t_2, \dots, t_m; [t_0, t_m]\big)/w_1(t_0); \tag{7.8}$$

$\wp_m(t_1, t_2, \dots, t_m | t_0)\Delta t_1 \Delta t_2 \dots \Delta t_m$ is the probability that, given a photoelectron emission occurs at time t_0, the next m emissions occur in the nonoverlapping intervals $[t_1, t_1 + \Delta t_1), [t_2, t_2 + \Delta t_2), \dots, [t_m, t_m + \Delta t_m)$, where $t_1 < t_2 < \dots < t_m$. The distribution of waiting times τ between a photoelectron emission at time t, and the next at time $t + \tau$, is

$$w(\tau|t) \equiv \wp_1(t+\tau|t) = \wp_2\big(t, t+\tau; [t, t+\tau]\big)/w_1(t). \tag{7.9}$$

For a stationary process $g^{(2)}(t, t+\tau)$ and $w(\tau|t)$ are independent of t.

We can now give a simple example of how (7.4) works. We will interpret this equation for the moment as a classical equation; therefore, the average is a classical average and the operator $\hat{I}(t)$ is read as the classical cycle averaged intensity (scaled to have units of photon flux) $\xi\bar{I}(t)$. Now (7.2) and (7.4)–(7.6) give

$$w_1(t) = \xi\langle\bar{I}(t)\rangle, \tag{7.10a}$$

$$w_2(t,t+\tau) = \xi^2\langle\bar{I}(t)\bar{I}(t+\tau)\rangle, \tag{7.10b}$$

and

$$\wp_2\big(t,t+\tau;[t,t+\tau]\big) = \xi^2\left\langle\bar{I}(t+\tau)\exp\left[-\xi\int_t^{t+\tau}dt'\bar{I}(t')\right]\bar{I}(t)\right\rangle. \tag{7.11}$$

For constant intensity light we then obtain

$$g^{(2)}(t,t+\tau) = 1, \tag{7.12a}$$

$$w(\tau|t) = \xi\bar{I}\exp(-\xi\bar{I}\tau). \tag{7.12b}$$

Equation (7.12b) reproduces (5.32); alternatively, our derivation of (5.32) provides an illustration, for a simplest case, of the derivation of (7.4).

We make one final observation about the relationship between nonexclusive and exclusive probability densities. We noted in Lecture 5 that for short enough times, apart from a scale factor, $g^{(2)}(t,t+\tau)$ and $w(\tau|t)$ are very nearly the same. The generalization of this result is clear from a comparison between (7.2) and (7.4). If each of the exponentials in (7.4) is replaced by unity, this expression becomes the same as (7.2). We can replace the exponentials by unity when the integrated flux over each of the intervals between the specified emission times is very small; that is to say, when the probability for additional photoelectron emissions to occur during these intervals is very small. This is often the case in photoelectron counting experiments, and is the reason why the second-order correlation function can be measured using a time-to-amplitude converter, which, more precisely, measures the distribution of waiting times [7.10, 7.11].

7.3 Quantum trajectories from the photoelectron counting distribution

Now to the question of how we might evaluate (7.4). There are a number of difficulties with this expression. First, it is an average taken over the state of the full system of source plus reservoir. Second, the average is to be evaluated with the operators written in normal order and time order, and they do not appear naturally ordered in that way in (7.4). Third, this is not a simple one-time average, it is a multi-time average which means that we must have some way of propagating the fields forwards in time.

In spite of these difficulties we are able to cast (7.4) into a manageable form for a wide class of systems. Actually, the expression we obtain will still generally be difficult to evaluate explicitly for anything other than the lowest values of m. But its form will suggest the path that takes us to the quantum trajectory formulation of the source dynamics. There is not time to go through all the details of the calculation, but the important points should be clear from an outline of what has to be done. Further detail is given by Carmichael et al. in their treatment of waiting times and atomic state reduction in resonance fluorescence [7.12]. Some mathematical points are also elucidated in Appendix A of the paper by Carmichael on shot noise and the spectrum of squeezing [7.13].

To begin with we decompose the field at the detector into free field and source field components (Sects. 1.4 and 2.4). We write

$$\hat{I}(t) = \eta[\hat{\mathcal{E}}_f^\dagger(t) + \hat{\mathcal{E}}_s^\dagger(t)][\hat{\mathcal{E}}_f(t) + \hat{\mathcal{E}}_s(t)], \tag{7.13}$$

where $\hat{\mathcal{E}}_f(t) \propto \hat{E}_f^{(+)}(t)$ and $\hat{\mathcal{E}}_s(t) \propto \hat{E}_s^{(+)}(t)$ are field operators written in photon flux units. Now if the reservoir is in the vacuum state, the free field operators will contribute nothing to the average in (7.4) because of the normal ordering and time ordering. Thus, in (7.4) we may use the substitution

$$\hat{I}(t) \to \eta\hat{\mathcal{E}}_s^\dagger(t)\hat{\mathcal{E}}_s(t). \tag{7.14}$$

The average is now taken over source operators alone; although, since these operators are evaluated at many different times the trace remains over the initial state of the source *and* the reservoir.

The next step is to take care of the operator ordering. We want to write the operator product inside the average as an explicitly ordered product; we now have to face the complexity that lies hidden in the double dots ::. The calculation in not difficult, but it is cumbersome to write down. We cannot order the operators until we expand the exponentials, which gives $m+1$ infinite sums and many products of integrals and field operators. The best thing to do is to first specialize to $m = 1$, or better still, consider the waiting-time distribution (7.9) which only involves one exponential. After seeing what must be done for a special case, it is not difficult to make the generalization to arbitrary m. The steps are as follows: (i) Expand the exponentials and write the field operators in explicit normal and time order. (ii) Write the resulting average as a trace over an expression written in superoperator form – for example, in evaluating the waiting-time distribution (7.9) the rewritten average is [7.12]

$$\mathrm{tr}\big[\mathcal{S}e^{L(\tau-\tau_k)}\mathcal{S}e^{L(\tau_k-\tau_{k-1})}\dots e^{L\tau_1}\mathcal{S}\hat{\chi}(t-r/c)\big], \tag{7.15}$$

where $\chi(t)$ is the density operator for the source plus reservoir, and L and $\mathcal{S}$ are defined by ($\hat{O}$ is any operator)

$$L\hat{O} = \frac{1}{i\hbar}[H, \hat{O}], \tag{7.16a}$$

$$\mathcal{S}\hat{O} = \hat{\mathcal{E}}_s(r/c)\hat{O}\hat{\mathcal{E}}_s^\dagger(r/c); \tag{7.16b}$$

H is the Hamiltonian (1.1), and the source field is evaluated at time $t = r/c$ so that the source operators themselves will be evaluated at $t = 0$ [Eqs. (1.60) and (2.61)]. (iii) Resum the sums of integrals that came from the exponentials using the identity

$$\begin{aligned}&\exp[(L + a\mathcal{S})x]\\&= \sum_{k=0}^{\infty} a^k \int_0^x dx_k \int_0^{x_k} dx_{k-1} \ldots \int_0^{x_2} dx_1\, e^{L(x-x_k)}\mathcal{S}e^{L(x_k-x_{k-1})}\mathcal{S}\ldots\mathcal{S}e^{Lx_1}.\end{aligned} \tag{7.17}$$

The result of all this is to replace (7.4) by

$$\begin{aligned}&\wp_m(t_1, t_2, \ldots, t_m; [t, t+T])\\&= \eta^m \mathrm{tr}\big[e^{(L-\eta\mathcal{S})(t+T-t_m)}\mathcal{S}\cdots\mathcal{S}e^{(L-\eta\mathcal{S})(t_2-t_1)}\mathcal{S}e^{(L-\eta\mathcal{S})(t_1-t)}\chi(t-r/c)\big].\end{aligned} \tag{7.18}$$

There is one more step to take before we have reached the result we want. In (7.18) the trace is taken over the combined system of source plus reservoir, and the superoperator L that appears in the propagator is defined in terms of the Hamiltonian (1.1) for the combined system. What we would like is to be able to evaluate a trace over the source alone. The basic ideas that allow us to accomplish this are contained in Sect. (3.2). Really (7.18) is just a complicated version of an equation like (3.19); it is a complicated multi-time average written in a formal superoperator language. Under the Born-Markoff assumption we can remove the trace over the reservoir as we did in deriving the master equation (Sect. 1.2) and the quantum regression theorem (Sect. 3.2). In doing this the superoperator L is replaced by the superoperator $\mathcal{L}$ that appears on the right-hand side of the master equation, and χ is replaced by the reduced density operator ρ:

$$\begin{aligned}&\wp_m(t_1, t_2, \ldots, t_m; [t, t+T])\\&= \eta^m \mathrm{tr}\big[e^{(\mathcal{L}-\eta\mathcal{S})(t+T-t_m)}\mathcal{S}\cdots\mathcal{S}e^{(\mathcal{L}-\eta\mathcal{S})(t_2-t_1)}\mathcal{S}e^{(\mathcal{L}-\eta\mathcal{S})(t_1-t)}\rho(t-r/c)\big].\end{aligned} \tag{7.19}$$

We now have the exclusive probability densities expressed as an average over source operators alone. From this expression the basic structure of the quantum trajectories is already visible.

Perhaps the best way to see this structure is to consider a ratio of two exclusive probability densities, which produces a conditional probability density. We write

$$\begin{aligned}\frac{\wp_{m+1}(t_1, t_2, \cdots, t_m, T; [t, t+T])}{\wp_m(t_1, t_2, \cdots, t_m; [t, t+T])} &= \mathrm{tr}[\mathcal{S}\rho_c(t+T-r/c)]\\&= \langle\hat{\mathcal{E}}_s^\dagger(t+T)\hat{\mathcal{E}}_s(t+T)\rangle_{\rho_c}.\end{aligned} \tag{7.20}$$

We have used the ratio on the left-hand side to define a density operator $\rho_c(t+T-r/c)$; thus,

$$\rho_c(t+T-r/c) = \frac{\bar{\rho}_c(t-r/c+T)}{\mathrm{tr}[\bar{\rho}_c(t-r/c+T)]}, \tag{7.21a}$$

where $\bar{\rho}_c(t+T-r/c)$ is the unnormalized operator

$$\begin{aligned}&\bar{\rho}_c(t+T-r/c)\\&= e^{(\mathcal{L}-\eta\mathcal{S})(t+T-t_m)}\mathcal{S}\cdots\mathcal{S}e^{(\mathcal{L}-\eta\mathcal{S})(t_2-t_1)}\mathcal{S}e^{(\mathcal{L}-\eta\mathcal{S})(t_1-t)}\rho(t-r/c).\end{aligned} \tag{7.21b}$$

Now the quantity on the right-hand side of (7.20) is the average of the source photon flux operator with respect to the density operator ρ_c. The quantity on the left-hand side gives the probability density for a photoelectron to be emitted at the time T, given that at $t-r/c$ the source density operator was $\rho(t-r/c)$, and given that a specified sequence of m photoelectron emissions (and no others) occurred at prescribed times during the interval between t and $t+T$. The relationship (7.20) then suggests that we interpret $\rho_c(t+T-r/c)$ as a *conditioned* source density operator – as the density operator for the source, given that at $t-r/c$ the source density operator was $\rho(t-r/c)$, and given that the specified sequence of m photoelectron emissions occurred at the prescribed times during the interval between t and $t+T$.

Now let us replace the reference to photoelectron emissions by a picture of photon emissions by the source. The density operator $\rho_c(t+T)$ depends on the quantum efficiency of the detector and must only describe the state of the source within the bounds of what is known from photoelectron emission sequences about the photons the source has emitted. To construct a visualization of source dynamics we should assume that every emitted photon is detected and replace η in (7.21b) by unity. We then extend the physical interpretation by noting that the bracket on the right-hand side of (7.21b) contains a product of propagators $e^{(\mathcal{L}-\mathcal{S})\Delta t}$ for the various intervals Δt between photon emissions, and m appearances of the superoperator $\mathcal{S}$. Reading this product from right to left the physical interpretation is as follows: the density operator evolves during the interval t_1-t when there are no photon emissions under the propagator $e^{(\mathcal{L}-\mathcal{S})(t_1-t)}$, collapses under the action of $\mathcal{S}$ at the time of the first emission, evolves during the next interval without photon emissions under the propagator $e^{(\mathcal{L}-\mathcal{S})(t_2-t_1)}$, collapses again under the action of $\mathcal{S}$, and so on. As we read we are generating a trajectory for ρ_c that takes this density operator from $\rho_c(t-r/c)=\rho(t-r/c)$ to the $\rho_c(t+T-r/c)$ defined by (7.20). The building blocks for constructing the trajectory are, first, two types of evolution – an evolution without photon emissions governed by the superoperator $(\mathcal{L}-\mathcal{S})$, and a collapse at the times of the photon emissions governed by the superoperator $\mathcal{S}$ – and, second, a specific set of times for the collapses (photon emissions). Since neither $\mathcal{S}$

nor $e^{(\mathcal{L}-\mathcal{S})\Delta t}$ preserve the density operator trace, the normalization will be introduced by hand, as in (7.21a).

The proposition is a little sketchy, but the sense is probably clear. To build a better understanding we will now approach the whole issue from a different direction.

7.4 Unravelling the master equation for the source

If the conditioned density operator ρ_c has meaning, what is its relationship to the density operator ρ that satisfies the source master equation? Remember, formally we write the source master equation as

$$\dot{\rho} = \mathcal{L}\rho. \tag{7.22}$$

Actually, this relationship is very easy to find; the calculation is much more direct than the one we have just discussed. The formal solution to (7.22) is

$$\rho(t) = e^{\mathcal{L}t}\rho(0). \tag{7.23}$$

We may add and subtract the superoperator $\mathcal{S}$ to $\mathcal{L}$, and use the identity (7.17), to obtain

$$\begin{aligned}\rho(t) &= e^{[(\mathcal{L}-\mathcal{S})+\mathcal{S}]t}\rho(0)\\ &= \sum_{m=0}^{\infty}\int_0^t dt_m \int_0^{t_m} dt_{m-1}\cdots\int_0^{t_2} dt_1\, e^{(\mathcal{L}-\mathcal{S})(t-t_m)}\mathcal{S}\\ &\quad\times e^{(\mathcal{L}-\mathcal{S})(t_m-t_{m-1})}\mathcal{S}\ldots\mathcal{S}e^{(\mathcal{L}-\mathcal{S})t_1}\rho(0).\end{aligned} \tag{7.24}$$

Now the quantity inside the integrals is the unnormalized conditioned density operator $\bar{\rho}_c(t)$ for an initial state $\rho_c(0) = \rho(0)$. We can interpret (7.24) as a generalized sum over all the photon emission pathways that the source might follow during its evolution from $t = 0$ to the time t. Each pathway may involve any number of photon emissions, from $m = 0$ up to $m = \infty$, and the times of the emissions can be any ordered sequence of m times in the interval $[0, t]$. What we are doing in defining a conditioned density operator is taking the quantity inside the integrals on the right-hand side of (7.24) out, normalizing it, and giving it a physical interpretation in terms of an evolution without photon emissions interrupted by collapses at the times of the photon emissions. At time t, for an initial state $\rho(0)$ and a particular sequence of photon emission times, the *conditioned source density operator* is given by

$$\rho_c(t) = \frac{\bar{\rho}_c(t)}{\mathrm{tr}[\bar{\rho}_c(t)]}, \tag{7.25a}$$

where $\bar{\rho}_c(t)$ is the unnormalized operator

$$\bar{\rho}_c(t) = e^{(\mathcal{L}-\mathcal{S})(t-t_m)}\mathcal{S}\cdots\mathcal{S}e^{(\mathcal{L}-\mathcal{S})(t_2-t_1)}\mathcal{S}e^{(\mathcal{L}-\mathcal{S})t_1}\rho(0). \tag{7.25b}$$

This procedure yields a decomposition of the quantum dynamics contained in the source master equation into an infinity of quantum paths, *quantum trajectories*, whose definition is based on separating the times at which photons materialize as photoelectrons at a detector (a conceptualized detector of unit quantum efficiency), from a quantum evolution over intervals of time during which photons, although watched for, are not materialized. The decomposition is something like a Feynman path integral [7.14]; although, with its basis in a master equation rather than a Schrödinger equation it is not precisely the same. We will refer to the quantum trajectories $\rho_c(t)$ as an *unravelling of the source dynamics* since it is a decomposition of the many tangled paths that the master equation (7.22) evolves forwards in time as a single package. From the development we have followed in this section it is probably clear that unravellings are not unique. We could choose any superoperator for $\mathcal{S}$. Of course, the photon emission picture is tied to the particular $\mathcal{S}$ define in (7.16b). But there are ways to look at the light radiated by a photoemissive source other than by direct photoelectron counting. These give different unravellings. We will say more about this in the next two lectures.

7.5 Stochastic wavefunctions

There is one piece missing from what we have seen so far. Equations (7.25) define a trajectory for a prescribed sequence of emission times. But the emission times of photoelectrons at a photodetector are random, and the emission times of the photons are surely random also. We have to build this randomness into the theory in a way that is statistically correct. We might note at this stage that photoelectron sequences are described by classical statistics. They are described within the language of classical stochastic processes. Corresponding to this, the randomness associated with the emission times that go into the integrand of (7.24) is simply a classical randomness. The peculiarly quantum-mechanical part of the density operator evolution occurs through the propagators $e^{(\mathcal{L}-\mathcal{S})\Delta t}$ and the action of $\mathcal{S}$, which only indirectly affects the determination of emission times. It is essentially (7.20) that determines when the emissions occur. More specifically, if the conditioned density operator at time t is $\rho_c(t)$, then the probability for an emission to occur in the interval $[t, t+\Delta t)$ is given by

$$p_c(t) = \mathrm{tr}[\mathcal{S}\rho_c(t)]\Delta t. \tag{7.26}$$

This is the product of the conditioned mean photon flux at time t and the time interval Δt.

Strictly, what we should do now is use the language of stochastic processes to define stochastic trajectories and show formally that these trajectories are statistically equivalent for calculating observed averages to the master equation (7.22). This is a laborious task that we do not have time for. Instead, we will define the stochastic quantum trajectories in an operational manner by using a numerical simulation to produce individual realizations. We then simply state the claimed statistical equivalence to the source master equation. Proof of this equivalence, or a strong indication that the formal proof can be done, will come from the examples treated in the following lectures.

Very often the form of the superoperators $(\mathcal{L} - \mathcal{S})$ and $\mathcal{S}$ allows the conditioned density operator $\rho_c(t)$ to be factorized as a pure state:

$$\rho_c(t) = |\psi_c(t)\rangle\langle\psi_c(t)|; \tag{7.27a}$$

we also write

$$\bar{\rho}_c(t) = |\bar{\psi}_c(t)\rangle\langle\bar{\psi}_c(t)|; \tag{7.27b}$$

explicit examples will be seen in the next lecture. Then the propagator $e^{(\mathcal{L}-\mathcal{S})\Delta t}$ for the density operator $\bar{\rho}_c(t)$ is replaced by a propagator for the state $|\bar{\psi}_c(t)\rangle$. Propagation without photon emission over a time Δt is given by

$$|\bar{\psi}_c(t+\Delta t)\rangle = e^{-(i/\hbar)H\Delta t}|\bar{\psi}_c(t)\rangle, \tag{7.28}$$

where H is a *non-Hermitian* Hamiltonian. At the time of a photon emission the unnormalized state undergoes a collapse

$$|\bar{\psi}_c(t)\rangle \to \hat{C}|\bar{\psi}_c(t)\rangle = \hat{\mathcal{E}}_s(r/c)|\bar{\psi}_c(t)\rangle. \tag{7.29}$$

Now our numerical simulation takes place over discrete time with a time step Δt. We obtain a stochastic trajectory for the conditioned wavefunction $|\psi_c(t_n)\rangle$, where $t_n = n\Delta t$. Given the wavefunction $|\psi_c(t_n)\rangle$, the wavefunction $|\psi_c(t_{n+1})\rangle$ is determined by the following algorithm: (i) Evaluate the collapse probability

$$p_c(t_n) = \langle\psi_c(t_n)|\hat{\mathcal{E}}_s^\dagger(r/c)\hat{\mathcal{E}}_s(r/c)|\psi_c(t_n)\rangle\Delta t. \tag{7.30}$$

(ii) Generate a random number r_n distributed uniformly on the interval $[0,1]$. (iii) Compare $p_c(t_n)$ with r_n and calculate $|\psi_c(t_{n+1})\rangle$ according to the rule

$$|\psi_c(t_{n+1})\rangle = \frac{\hat{C}|\psi_c(t_n)\rangle}{\sqrt{\langle\psi_c(t_n)|\hat{C}^\dagger\hat{C}|\psi_c(t_n)\rangle}} \qquad p_c(t_n) \le r_n, \tag{7.31a}$$

$$|\psi_c(t_{n+1})\rangle = \frac{e^{-(i/\hbar)H\Delta t}|\psi_c(t_n)\rangle}{\sqrt{\langle\psi_c(t_n)|e^{(i/\hbar)(H^\dagger-H)\Delta t}|\psi_c(t_n)\rangle}} \qquad p_c(t_n) > r_n. \tag{7.31b}$$

The result of all of this is a stochastic quantum mapping between the times t_m (separated by many Δt) at which the collapses occur:

$$|\psi_c(t_{m+1})\rangle = \frac{\hat{C}e^{-(i/\hbar)H\tau_{m+1}}|\psi_c(t_m)\rangle}{\sqrt{\langle\psi_c(t_m)|e^{(i/\hbar)H^\dagger\tau_{m+1}}\hat{C}^\dagger\hat{C}e^{-(i/\hbar)H\tau_{m+1}}|\psi_c(t_m)\rangle}}, \quad (7.32)$$

where $\tau_{m+1} = t_{m+1} - t_m$ is a random time whose statistics depend on the stochastic wavefunction itself through the relationship (7.30). The numerical algorithm incorporates these statistics "on the fly" by taking many infinitesimal steps Δt. In this way we do not need an explicit distribution for τ_{m+1}. In anything other than the simplest examples this distribution would be very difficult to calculate. It is a waiting-time distribution; but it is conditioned on the times of the m preceding collapses. In fact, this distribution is determined by a ratio of exclusive probability densities like the one given by (7.20). We can obtain it in closed form if we can calculate all the exclusive probability densities in closed form. Resonance fluorescence provides an example where this is possible because the emission sequences are Markoffian, and therefore the exclusive probability densities factorize [7.12]. Note that (7.32) assumes H does not depend explicitly on time. When this is not the case the generalization is obvious.

The claim is that the quantum trajectories generated in this way are statistically equivalent to the standard solution to the source master equation. For example, if an ensemble of such trajectories is generated starting in the same initial state, then the ensemble average of $\rho_c(t)$ is the density operator $\rho(t)$. Note that when $\rho(t)$ comes to a steady state, $\rho_c(t)$ will not; each trajectory keeps up its stochastic evolution. The trajectory is, however, stationary in a statistical sense, and the steady-state result for ρ can be calculated as a time average of $\rho_c(t)$. Other ensemble and time averages can be calculated that correspond to various observed quantities, such as the mean intensity of the source or the photoelectron counting distribution. We will see some examples in the remaining lectures.

References

[7.1] C. W. Gardiner and M. J. Collett, *Phys. Rev.* A **31**, 3761 (1985).
[7.2] E. B. Davies, *Quantum Theory of Open Systems*, Academic Press: London, 1976.
[7.3] M. D. Srinivas and E. B. Davies, *Optica Acta* **28**, 981 (1981).
[7.4] L. Mandel, *Optica Acta* **28**, 1447 (1981).
[7.5] M. D. Srinivas and E. B. Davies, *Optica Acta* **29**, 235 (1982).
[7.6] P. L. Kelly and W. H. Kleiner, *Phys. Rev.* **136**, A316 (1964).
[7.7] R. J. Glauber, *Phys. Rev.* **130**, 2529 (1963).
[7.8] R. J. Glauber, *Phys. Rev.* **131**, 2766 (1963).

[7.9] B. Saleh, *Photoelectron Statistics*, Springer: Berlin, 1978, Chap. 3.
[7.10] F. Davidson and L. Mandel, *J. Appl. Phys.* **39**, 62 (1968).
[7.11] H. J. Kimble, M. Dagenais, and L. Mandel, *Phys. Rev. Lett.* **39**, 691 (1977).
[7.12] H. J. Carmichael, S. Singh, R. Vyas, and P. R. Rice, *Phys. Rev.* A **39**, 1200 (1989).
[7.13] H. J. Carmichael, *J. Opt. Soc. Am.* B **4**, 1588 (1987).
[7.14] R. P. Feynman, *Reviews of Modern Physics* **20**, 367 (1948).

Lecture 8 – Quantum Trajectories II

We have suggested that the operator master equation for a photoemissive source is statistically equivalent to a stochastic quantum mapping. Each iteration of the mapping involves a quantum evolution under a nonunitary Schrödinger equation, for a random interval of time, followed by a wavefunction collapse at the end of this interval. In general, the probability distribution governing the duration of the quantum evolution depends on the past history of the source. In most cases it will be very difficult to implement this mapping analytically. However, it is quite easy to implement on a computer. The computer simulations generate "trajectories" for a stochastic wavefunction that describes the current state of the quantum-mechanical source, conditioned on a particular past history of coherent evolution and collapse. Time series obtained from these trajectories have a direct statistical correspondence to the fluctuating signals obtained by monitoring a single quantum system (not an ensemble) in the laboratory. They can be analyzed like experimental data – for a stationary process, by averaging in time; the time averages reproduce the usual quantum-mechanical average.

We now apply this quantum trajectory method to various elementary examples, and show that it reproduces results obtained by conventional methods. The material presented in this lecture is taken from a presentation by Carmichael and Tian at the 1990 Annual Meeting of the Optical Society of America [8.1].

8.1 Damped atoms and cavities

Perhaps the simplest example we can consider is spontaneous emission from a two-state atom. In this example the picture obtained from the quantum trajectory approach is a picture that has been presented in many guises before. It is the picture of jumps between discrete atomic states inherent in the Einstein rate equations [Eqs. (3.4a) and (3.4b)]. A closely related example is the decay of an optical cavity mode prepared in a Fock state. We will look first at the atomic example and then at the decaying cavity mode.

We consider a single two-state atom (lower state $|1\rangle$ and upper state $|2\rangle$) described by the source master equation (2.26) (with $\bar{n} = 0$). The field radiated by the atom is given in terms of source operators by (2.61). To make things as simple as possible we will assume that the detector sees the complete 4π solid angle into which the photon is emitted. The source field operator scaled to give photon flux into the detector is then

$$\hat{\mathcal{E}}_s(t) = \sqrt{\gamma}\sigma_-(t - r/c), \tag{8.1}$$

where γ is the Einstein A coefficient and r is the distance from the source to the detector; the overall phase of this field is unimportant since the decomposition of the master equation dynamics we consider is based on intensity. The superoperators $(\mathcal{L} - \mathcal{S})$ and $\mathcal{S}$ that govern the coherent evolution and collapse, respectively, are defined by the relationships

$$\mathcal{S}\bar{\rho}_c = \gamma\sigma_-\bar{\rho}_c\sigma_+, \tag{8.2a}$$

$$(\mathcal{L} - \mathcal{S})\bar{\rho}_c = -i\tfrac{1}{2}\omega_A[\sigma_z, \bar{\rho}_c] - \frac{\gamma}{2}(\sigma_+\sigma_-\bar{\rho}_c + \bar{\rho}_c\sigma_+\sigma_-), \tag{8.2b}$$

where $\bar{\rho}_c$ is the *unnormalized* conditioned density operator – the density operator for the atom conditioned on its past. In this example the conditioned density operator may be written in terms of a pure state wavefunction:

$$\rho_c(t) = |\psi_c(t)\rangle\langle\psi_c(t)|. \tag{8.3}$$

The dynamical evolution of the unnormalized wavefunction $|\bar{\psi}_c(t)\rangle$ is governed by the nonunitary Schrödinger equation

$$\frac{d}{dt}|\bar{\psi}_c\rangle = \frac{1}{i\hbar}H|\bar{\psi}_c\rangle, \tag{8.4a}$$

with the non-Hermitian Hamiltonian

$$H = \tfrac{1}{2}\hbar\omega_A\sigma_z - i\hbar\frac{\gamma}{2}\sigma_+\sigma_-. \tag{8.4b}$$

The evolution generated by (8.4a) is interrupted by collapses

$$|\bar{\psi}_c\rangle \to \hat{C}|\bar{\psi}_c\rangle, \tag{8.5a}$$

with collapse operator

$$\hat{C} = \sqrt{\gamma}\sigma_-. \tag{8.5b}$$

The probability for a collapse to occur in the interval $(t, t + \Delta t]$ is given by

$$\begin{aligned} p_c(t) &= \mathrm{tr}[\mathcal{S}\rho_c(t)]\Delta t \\ &= (\gamma\Delta t)\langle\psi_c(t)|\sigma_+\sigma_-|\psi_c(t)\rangle \\ &= (\gamma\Delta t)\frac{\langle\bar{\psi}_c(t)|\sigma_+\sigma_-|\bar{\psi}_c(t)\rangle}{\langle\bar{\psi}_c(t)|\bar{\psi}_c(t)\rangle}. \end{aligned} \tag{8.6}$$

The spontaneous emission example is sufficiently simple that we can actually solve the trajectory equations (8.4a) and (8.5a) analytically. Assume an arbitrary initial condition

$$|\bar{\psi}_c(0)\rangle \equiv |\psi_c(0)\rangle = c_1(0)|1\rangle + c_2(0)|2\rangle. \tag{8.7}$$

From (8.4a) and (8.4b) we find that the unnormalized amplitudes $\bar{c}_1(t)$ and $\bar{c}_2(t)$ obey the equations

$$\dot{\bar{c}}_1 = \tfrac{1}{2}i\omega_A\bar{c}_1, \tag{8.8a}$$

$$\dot{\bar{c}}_2 = -(\gamma/2 + \tfrac{1}{2}i\omega_A)\bar{c}_2. \tag{8.8b}$$

The solutions are

$$\bar{c}_1(t) = c_1(0)e^{\frac{1}{2}i\omega_A t}. \tag{8.9a}$$

$$\bar{c}_2(t) = c_2(0)e^{-(\gamma/2)t}e^{-\frac{1}{2}i\omega_A t}. \tag{8.9b}$$

The normalized amplitudes are then

$$c_1(t) = \frac{c_1(0)}{\sqrt{|c_1(0)|^2 + |c_2(0)|^2e^{-\gamma t}}}e^{\frac{1}{2}i\omega_A t}, \tag{8.10a}$$

$$c_2(t) = \frac{c_2(0)e^{-(\gamma/2)t}}{\sqrt{|c_1(0)|^2 + |c_2(0)|^2e^{-\gamma t}}}e^{-\frac{1}{2}i\omega_A t}. \tag{8.10b}$$

Equations (8.10) provide the solution for the conditioned wavefunction during the coherent evolution that occurs between collapses:

$$\begin{aligned}|\psi_c(t)\rangle &= c_1(t)|1\rangle + c_2(t)|2\rangle \\ &= \frac{c_1(0)e^{\frac{1}{2}i\omega_A t}|1\rangle + c_2(0)e^{-(\gamma/2)t}e^{-\frac{1}{2}i\omega_A t}|2\rangle}{\sqrt{|c_1(0)|^2 + |c_2(0)|^2e^{-\gamma t}}}.\end{aligned} \tag{8.11}$$

The probability for a collapse during $(t, t + \Delta T]$ is given by

$$p_c(t) = (\gamma\Delta t)\frac{|c_2(0)|^2e^{-\gamma t}}{|c_1(0)|^2 + |c_2(0)|^2e^{-\gamma t}}; \tag{8.12}$$

for an initially excited atom ($c_1(0) = 0$) this probability is independent of time. Clearly there is only one collapse in each trajectory since (8.5a) and (8.5b), and (8.9a) and (8.9b) give (after normalizing the states before and after the collapse)

$$|\psi_c(t)\rangle = c_1(t)|1\rangle + c_2(t)|2\rangle \to |1\rangle. \tag{8.13}$$

Once the atom reaches the lower state $|1\rangle$ the nonunitary Schrödinger equation [solutions (8.10)] simply keeps it there forever; obviously, there can be one and only one photon emission from a single undriven atom.

From the solution (8.11) we can get some sense of what the *conditioned* wavefunction means. Equation (8.11) gives the state of the atom conditioned on the fact that it has not yet emitted a photon; it is the state of the atom before it collapses. We find then that if $c_1(0) \neq 0$ this state approaches $|1\rangle$ for times much longer that the lifetime γ^{-1}. What this tells us is that if we have waited many lifetimes without seeing a photon emission, it is very likely that the atom actually began in the lower state $|1\rangle$, from which it could not emit . Thus, in waiting for a photon that never came we gain the information that the atom must be in the lower state; therefore, the atom reaches the lower state either by a collapse and photon emission [Eq. (8.13)], or by eventually convincing us that it was actually in the lower state all the time.

An atom prepared in the upper state must collapse into the lower state. A sample trajectory for the conditioned wavefunction is defined by a function $c_2(t)$, that starts with $c_2(0) = 1$, and remains constant until some random time at which it switches to the value $c_2(t) = 0$, remaining there forever; similarly, the function $c_1(t)$ starts with $c_1(0) = 0$ and switches up to the value $c_1(t) = 1$, remaining there forever. This is the jump that we all expect as the atom emits its quantum of energy. The time of emission for each quantum trajectory is random; in the computer it is determined by comparing a random number with the collapse probability (8.12) at each step of the stochastic simulation, as described in Sect. (7.5). If a large number of these emissions is simulated and the number of emissions occurring in $(t, t + \Delta t]$ is plotted against t, we recover the exponential decay illustrated in Fig. 8.1. This corresponds to the exponential decay obtained from the emission probability $(\gamma \Delta t)\rho_{22}(t)$, where $\rho_{22}(t) = e^{-\gamma t}$ is the solution to the Einstein rate equations.

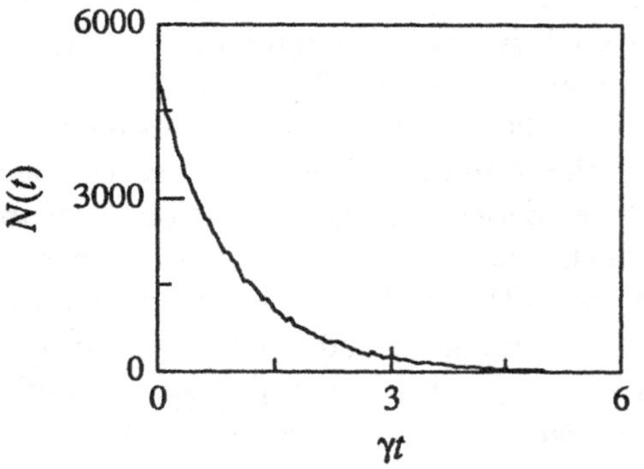

Fig. 8.1. Number of emissions in the interval γt to $\gamma(t + \Delta t)$ versus γt for a simulation of 100,000 spontaneous emission trajectories ($\gamma \Delta t = 0.05$).

The extension of these ideas to the decay of a cavity mode prepared in a Fock state is probably fairly obvious. In this case the operator master equation for the source is (1.47) and the relationship between the radiated field and source operators is given in (1.60). If the detector intercepts the entire cavity output beam, the source field scaled to give photon flux into the detector is

$$\hat{\mathcal{E}}_s(t) = \sqrt{2\kappa}\, a(t - r/c). \tag{8.14}$$

In place of (8.2a) and (8.2b) we have

$$\mathcal{S}\bar{\rho}_c = 2\kappa a \bar{\rho}_c a^\dagger, \tag{8.15a}$$

$$(\mathcal{L} - \mathcal{S})\bar{\rho}_c = -i\omega_C[a^\dagger a \bar{\rho}_c] - \kappa(a^\dagger a \bar{\rho}_c + \bar{\rho}_c a^\dagger a). \tag{8.15b}$$

Once again, the conditioned density operator factorizes as a pure state and satisfies the nonunitary Schrödinger equation (8.4a). The non-Hermitian Hamiltonian is

$$H = \hbar\omega_C a^\dagger a - i\hbar\kappa a^\dagger a. \tag{8.16}$$

The collapse (8.5a) is governed by the collapse operator

$$\hat{C} = \sqrt{2\kappa}\, a, \tag{8.17}$$

and the collapse probability is given by

$$\begin{aligned} p_c(t) &= (2\kappa\Delta t)\mathrm{tr}[\mathcal{S}\rho_c(t)] \\ &= (2\kappa\Delta t)\langle\psi_c(t)|a^\dagger a|\psi_c(t)\rangle \\ &= (2\kappa\Delta t)\frac{\langle\bar{\psi}_c(t)|a^\dagger a|\bar{\psi}_c(t)\rangle}{\langle\bar{\psi}_c(t)|\bar{\psi}_c(t)\rangle}. \end{aligned} \tag{8.18}$$

It is again possible to solve the evolution between collapses analytically. We will not bother with the details. The main point is that the amplitude equations are uncoupled as they are in (8.8a) and (8.8b); consequently, if the cavity mode is in a Fock state, it remains in that Fock state until the next collapse (photon emission) occurs. At that time the effect of the collapse operator (8.17) is to take the Fock state $|n\rangle$ to the Fock state $|n-1\rangle$. Clearly, an initial state $|N\rangle$ will undergo N jumps, at N random times, until the cavity mode reaches the vacuum state, where it will remain forever. A sample trajectory is illustrated in Fig. 8.2(a). On average the dwell time in each Fock state becomes longer as the level of excitation decreases; this is because the collapse probability (8.18) depends on the conditioned mean photon flux $\sqrt{2\kappa}\,\langle\psi_c(t)|a^\dagger a|\psi_c(t)\rangle$ which decreases as the system descends the random staircase. Figure 8.2(b) shows the evolution of the average intracavity photon number, calculated by averaging 10,000 realizations of the conditioned mean photon number $\langle a^\dagger a\rangle_c \equiv \langle\psi_c(t)|a^\dagger a|\psi_c(t)\rangle$. The ensemble average over trajectories shows the exponential decay given by (3.3).

8.2 Resonance fluorescence

Both of the examples we have just seen are really rather trivial. The quantum trajectories for both are elementary examples of Markoff processes on discrete state spaces. Anyone who is familiar with Markoff processes and a little quantum mechanics could have concocted simulations to produce the quantum trajectories shown in Figs. 8.1 and 8.2. But we have something more than a concoction. We have a well-defined formal procedure for constructing the stochastic process from an operator master equation. In general the quantum dynamics for a given source will not be as transparent as in the foregoing examples, and the "concoction" approach will not work.

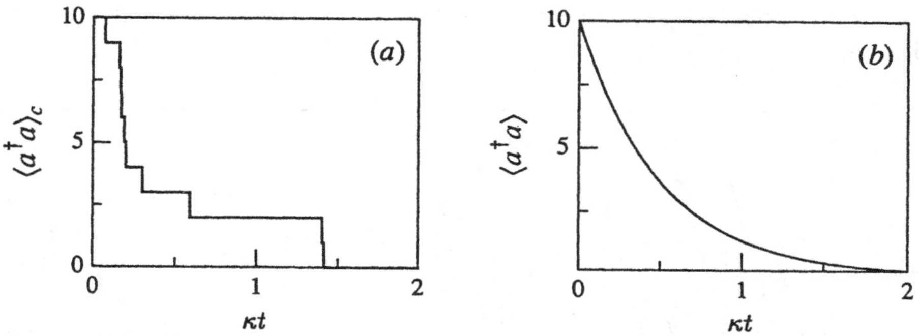

Fig. 8.2. (a) Sample quantum trajectory showing the conditioned mean photon number for a damped cavity mode prepared in the Fock state $|10\rangle$. (b) Average of the conditioned mean photon number for 10,000 trajectories.

The first such nontrivial example we look at is resonance fluorescence. The discussion that follows is an extension of work by Carmichael et al. [8.2].

To model resonance fluorescence the master equation for the atomic source changes from (2.26) to (2.62); we add the dipole interaction with the coherent driving field, proportional to the Rabi frequency Ω. If we keep the assumption that the detector sees all the fluorescence, the source field in photon number units is still (8.1). The collapse of the atomic state is still described by the superoperator relation (8.2a), and (8.2b) changes to

$$\begin{aligned}(\mathcal{L}-\mathcal{S})\bar{\rho}_c = &-i\tfrac{1}{2}\omega_A[\sigma_z,\bar{\rho}_c] - i(\Omega/2)[e^{-i\omega_A t}\sigma_+ + e^{i\omega_A t}\sigma_-,\bar{\rho}_c] \\ &-\frac{\gamma}{2}(\sigma_+\sigma_-\bar{\rho}_c + \bar{\rho}_c\sigma_+\sigma_-).\end{aligned} \tag{8.19}$$

The rest of the formulation outlined in (8.1)–(8.6) is the same, with the Hamiltonian (8.4b) changed to

$$H = \tfrac{1}{2}\hbar\omega_A\sigma_z + \hbar(\Omega/2)[e^{-i\omega_A t}\sigma_+ + e^{i\omega_A t}\sigma_-] - i\hbar\frac{\gamma}{2}\sigma_+\sigma_-. \tag{8.20}$$

Now from our previous discussion of resonance fluorescence we know that a single fluorescing atom evolves to a stationary state. In conventional language the density operator for the stationary state is defined by (3.64a) and (3.64b). In the quantum trajectory approach we would expect the evolution of the conditioned wavefunction to be governed by a stationary stochastic process. The stochastic process is, in fact, still fairly simple because the collapse relation (8.13) still applies. Thus, after each collapse (photon emission) the atom is in its lower state; this means that the evolution between collapses is always solved from the same initial condition. Unlike the spontaneous emission example, in the presence of the driving field the atom does not remain in the lower state after a collapse; rather, it evolves to a new state $|\psi_c(t)\rangle = c_1(t)|1\rangle + c_2(t)|2\rangle$ with $c_2(t) \neq 0$, where t is now the time since the previous collapse. In this way the atom continuously generates a nonzero probability for making a further collapse and emitting another photon.

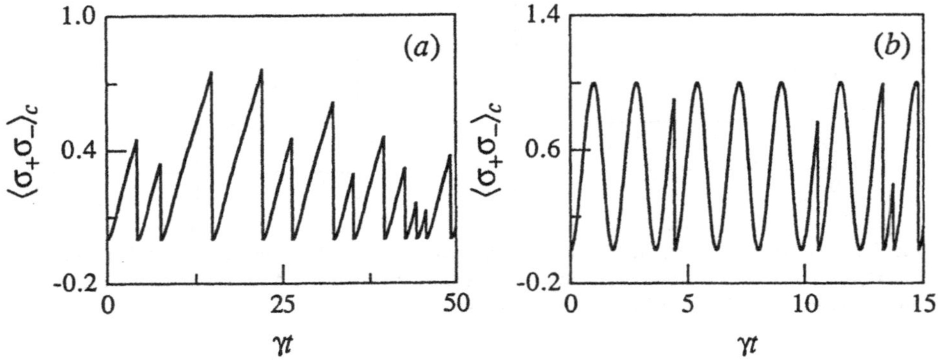

Fig. 8.3. (a) Sample quantum trajectories showing the conditioned upper state probability of an atom undergoing resonance fluorescence. (a) Weak excitation, $\Omega/\gamma = 0.7$; (b) strong excitation, $\Omega/\gamma = 3.5$.

The equations obeyed by the unnormalized amplitudes during the coherent evolution are minor variations of (8.8a) and (8.8b):

$$\dot{\bar{c}}_1 = \tfrac{1}{2}i\omega_A\bar{c}_1 + i(\Omega/2)e^{i\omega_A t}\bar{c}_2, \tag{8.21a}$$

$$\dot{\bar{c}}_2 = -(\gamma/2 + \tfrac{1}{2}i\omega_A)\bar{c}_2 + i(\Omega/2)e^{-i\omega_A t}\bar{c}_1. \tag{8.21b}$$

For an initial state $|\psi_c(0)\rangle = |1\rangle$ the solutions to these equations give the unnormalized amplitudes

$$\bar{c}_1(t) = e^{-(\gamma/4)t}e^{\frac{1}{2}i\omega_A t}\left[\cosh(\delta t) + \frac{(\gamma/2)}{2\delta}\sinh(\delta t)\right], \tag{8.22a}$$

$$\bar{c}_2(t) = ie^{-(\gamma/4)t}e^{-\frac{1}{2}i\omega_A t}\frac{\Omega}{2\delta}\sinh(\delta t), \tag{8.22b}$$

where

$$2\delta = \sqrt{(\gamma/2)^2 - \Omega^2}. \tag{8.23}$$

The collapse probability in the time interval $(t, t+\Delta t]$ is then given by

$$p_c(t) = (\gamma\Delta t)|c_2(t)|^2 = (\gamma\Delta t)\frac{|\bar{c}_2(t)|^2}{|\bar{c}_1(t)|^2 + |\bar{c}_2(t)|^2}. \tag{8.24}$$

Figure 8.3 shows two examples of quantum trajectories for resonance fluorescence. The full quantum state could be represented by a stochastic motion on the Bloch sphere; in Fig. 8.3 the upper state probability $|c_2(t)|^2$ is plotted. The vertical jumps return the atom to the lower state at the times of the photon emissions; these are the collapses responsible for photon antibunching in resonance fluorescence (Sect. 3.5). Notice that for strong excitation [Fig. 8.3(b)] coherent Rabi oscillations occur between the emissions.

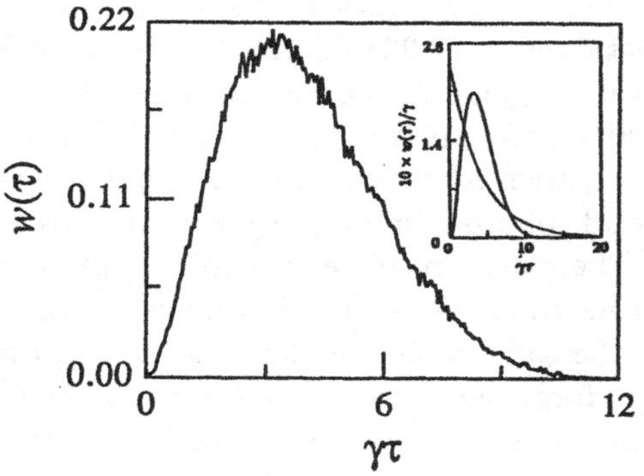

Fig. 8.4. Waiting-time distribution for resonance fluorescence obtained from a histogram of the time intervals between collapses (photon emissions) in the simulation of Fig. 8.3(a). The inset shows the distribution calculated analytically in [8.2].

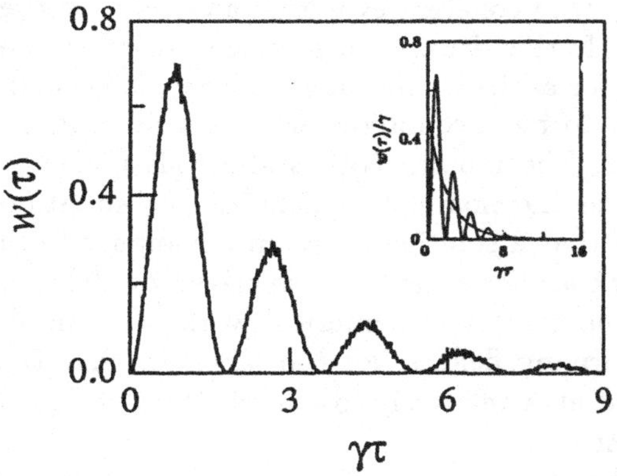

Fig. 8.5. Waiting-time distribution for resonance fluorescence obtained from a histogram of the time intervals between collapses (photon emissions) in the simulation of Fig. 8.3(b). The inset shows the distribution calculated analytically in [8.2].

From simulations like those illustrated in Fig. 8.3 it is possible to carry out photoelectric counting experiments in the computer. We simply count the number of collapses that occur in a counting time T. By repeating the process for many counting intervals we build up a histogram of the number of counting intervals that produce n photoelectron counts. The normalized histogram is the photoelectron counting distribution. We can also obtain waiting-time distributions in an equivalent manner. Figures 8.4 and 8.5 show two examples of waiting-time distributions obtained from quantum trajectories for resonance fluorescence. For comparison the inset shows the waiting-time distribution calculated analytically in [8.2]. The agreement is very good. Of course, the numerical simulations show residual sampling fluctuations, much like those expected in a laboratory experiment.

8.3 Cavity mode driven by thermal light

For an example like resonance fluorescence, where everything needed to simulate the quantum trajectories is contained in (8.22)–(8.24), the numerical simulations are very efficient. However, in general, the numerical work can be increased by a number of factors. First, often it is not possible to solve for the conditioned state $|\psi_c(t)\rangle$ explicitly; then a numerical differential equation solver must do this for us. Second, photon emission sequences in resonance fluorescence are Markoffian. The emission sequences are completely specified by the distribution of waiting times between adjacent emissions. This is because the atom returns to the same state, the lower state $|1\rangle$, on every collapse. After it does this it has forgotten all about where it has been in the past. More generally, each time the source collapses it collapses to a different state. The collapsed state depends on the state before the collapse, which in turn depends on the history of coherent evolution and collapse the source has experienced in the past. In this situation a general solution to the nonunitary Schrödinger equation, for arbitrary initial conditions, is needed.

These complications are likely to be encountered when considering an optical cavity mode as the source. The infinite Fock state basis makes it unlikely that a general solution to the nonunitary Schrödinger equation can be found, and even less likely that a solution exists in a compact form suitable for fast numerics. We now consider a cavity mode driven by thermal light. This is an example where the additional numerical work is required. However, if the intensity of the driving field is not too large, so that the Fock state basis can be truncated at a relatively low level, the numerical requirements are still quite modest.

Thermal excitation adds another complication. Since it is incoherent we are not able to factorize the conditioned density operator as a pure state. Equation (8.15a) holds for describing the collapse. But (8.15b) is replaced by

$$\begin{aligned}(\mathcal{L}-\mathcal{S})\bar{\rho}_c = & -i\omega_C[a^\dagger a\bar{\rho}_c] - \kappa(a^\dagger a\bar{\rho}_c + \bar{\rho}_c a^\dagger a) \\ & + 2\kappa\bar{n}(a\bar{\rho}_c a^\dagger + a^\dagger \rho_c a - a^\dagger a\rho_c - \rho_c a^\dagger a);\end{aligned} \tag{8.25}$$

the term proportional to $\bar{n}$ does not allow us to use a pure state for describing the evolution between collapses. Nevertheless, the general formalism still holds; it just has to be implemented in density matrix form, with the collapse probability for the interval $(t, t+\Delta t]$ given by

$$p_c(t) = \mathrm{tr}[\mathcal{S}\rho_c(t)]\Delta t = (2\kappa\Delta t)\mathrm{tr}[\rho_c(t)a^\dagger a]. \tag{8.26}$$

Figure 8.6 shows results for $\bar{n} = 1$. The thermal light is turned on at $t = 0$ and the figure shows the transient behavior as the cavity mode approaches a stationary state. Figure 8.6(a) shows a sample quantum trajectory for the conditioned mean photon number $\mathrm{tr}[\rho_c(t)a^\dagger a]$; Fig. 8.6(b) is the average of 10,000 such trajectories and reproduces the exponential filling of the cavity described by the conventional mean-value equation (3.3). Examples of trajectories for higher intensity light are shown in Fig. (8.7).

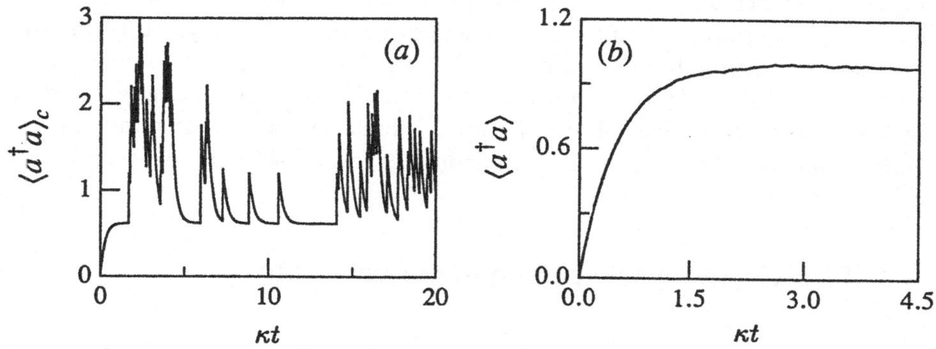

Fig. 8.6. (a) Sample quantum trajectory showing the conditioned mean photon number for a cavity driven by thermal light. The thermal light turns on at $t = 0$ and injects a photon flux $2\kappa\bar{n} = 2\kappa$ ($\bar{n} = 1$). The Fock state basis is truncated at 20 photons. (b) Ensemble average of 10,000 such trajectories.

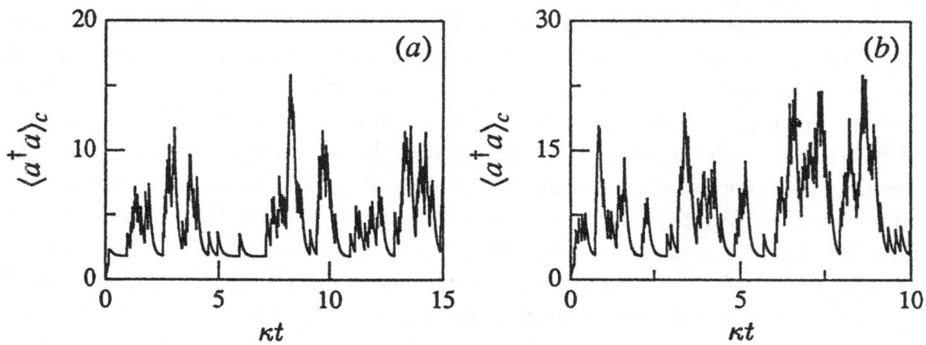

Fig. 8.7. Sample quantum trajectories showing the conditioned mean photon number for a cavity driven by thermal light. (a) The thermal light turns on at $t = 0$ and injects a photon flux $2\kappa\bar{n} = 10\kappa$ ($\bar{n} = 5$). The Fock state basis is truncated at 50 photons. (b) The thermal light turns on at $t = 0$ and injects a photon flux $2\kappa\bar{n} = 20\kappa$ ($\bar{n} = 10$). The Fock state basis is truncated at 80 photons.

These trajectories show a surprising feature that tells us a little more about the nature of the conditioned quantum state. The sudden jumps in the conditioned mean photon number occur when the state collapses as a photon is emitted from the cavity. But the jumps are upwards, not downwards as in Fig. 8.2. How can the emission of a photon make the number of photons in the cavity increase? The explanation is that the conditioned mean photon number is the mean of $a^\dagger a$ with respect to a state that is conditioned on everything that has taken place along the trajectory in the past. Every twist of this trajectory adds information to the memory. The conditioned mean photon number propagates information; it is not an actual photon number *out there* in the cavity. For a thermal state the observation of one collapse, one photon emitted, means another is very likely, at twice the average rate, immediately following the first. Thus, the photon bunching of thermal light (Sect. 3.4) is built into the conditioned state as *upwards jumps* in the conditioned mean photon number, which gives upwards jumps in the collapse probability [Eq. (8.26)] immediately following each collapse.

8.4 The degenerate parametric oscillator

Lecture 6 was devoted to the homodyne detection of squeezed light. In the next lecture we will see how the quantum trajectory approach can be used to treat homodyne detection. But first, let us look at squeezed light by direct photoelectric detection. The source master equation is based on the master equation (2.63) for the degenerate parametric oscillator. However, we will not take this master equation directly as it is written. We are interested in below threshold operation, where the quantum-classical correspondence led us to the Fokker-Planck equations (4.72) and (4.73). In these equations the coupling between fluctuations in the pump mode and the subharmonic mode has disappeared; the pump field simply enters the Fokker-Planck equation for the subharmonic mode through the parameter λ. We can build this simplification into the master equation directly. Essentially, we assume that the density operator ρ factorizes into a product of density operators for the two cavity modes. We then write a master equation for each. The density operator for the pump mode satisfies the master equation for a cavity driven by the coherent field $\bar{\mathcal{E}}_i$ – the second, fourth, and sixth terms on the right-hand side of (2.63); the master equation for the subharmonic mode is obtained from the first, third, and fifth terms on the right-hand side of (2.63), with the coherent state amplitude of the pump substituted for the operator b:

$$\begin{aligned}\dot{\rho} = &-i\omega_C[a^\dagger a, \rho] + (\kappa\lambda/2)[a^{\dagger 2}e^{-i2\omega_C t} - a^2 e^{i2\omega_C t}, \rho] \\ &+ \kappa(2a\rho a^\dagger - a^\dagger a\rho - \rho a^\dagger a).\end{aligned} \tag{8.27}$$

Here λ is the pump parameter defined below (4.64).

Now the superoperator governing the collapse is defined by (8.15a) and the coherent evolution between collapses is governed by

$$\begin{aligned}(\mathcal{L} - \mathcal{S})\bar{\rho}_c = &-i\omega_C[a^\dagger a, \bar{\rho}_c] + (\kappa\lambda/2)[a^{\dagger 2}e^{-i2\omega_C t} - a^2 e^{i2\omega_C t}, \bar{\rho}_c] \\ &- \kappa(a^\dagger a\bar{\rho}_c + \bar{\rho}_c a^\dagger a).\end{aligned} \tag{8.28}$$

It is again possible to factorize $\bar{\rho}_c$ as a pure state and use the nonunitary Schrödinger equation (8.4a). The non-Hermitian Hamiltonian is

$$H = \hbar\omega_C a^\dagger a + i\hbar(\kappa\lambda/2)(a^{\dagger 2} e^{-i2\omega_C t} - a^2 e^{i2\omega_C t}) - i\hbar\kappa a^\dagger a. \tag{8.29}$$

The collapse probability for the interval $(t, t + \Delta t]$ is calculated from (8.18).

A sample quantum trajectory for the conditioned mean photon number in the subharmonic mode is shown in Fig. 8.8(a). Figure 8.8(b) is the average of 10,000 such trajectories and shows the build-up of the mean photon number in the cavity after the pump is turned on at $t = 0$. Note how, once again, the collapse can cause an *upwards* jump in the conditioned mean photon number. In this example some of the jumps are upwards and some are downward. The reason for this is that photons are created in pairs inside the cavity. When the first photon of a pair is emitted from the cavity the conditioned mean photon number, and hence the collapse probability (8.18), makes an upwards jump; this ensures that the second photon will be emitted within a short time $[\sim (2\kappa)^{-1}]$ after the first. After the second photon has been emitted the collapse decreases the conditioned mean photon number, which in a few cavity lifetimes returns to its steady-state value.

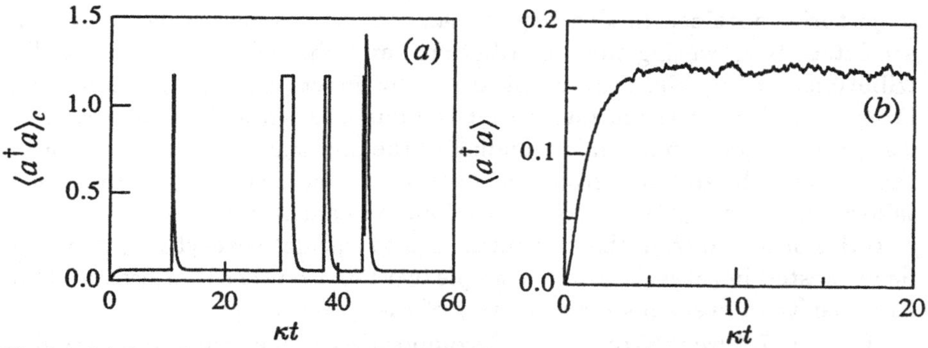

Fig. 8.8. (a) Sample quantum trajectory showing the conditioned mean photon number for a degenerate parametric oscillator operated 50% below threshold ($\lambda = 0.5$). The pump light is turned on a $t = 0$. The Fock state basis is truncated at 10 photons. (b) Ensemble average of 10,000 such trajectories.

The pairing of photon emissions leads to an imbalance between even and odd numbers of photoelectron counts in the photoelectron counting distribution. We have already mentioned this in Sect. 6.5. Figure 8.9 shows a photoelectron counting distribution obtained by counting the collapses (photon emissions) for many quantum trajectories of the sort illustrated in Fig. 8.8(a). The even-odd oscillations are large. The inset shows the distribution obtained by Wolinsky and Carmichael [8.3] for the same parameters, using a related but quite different method. This photoelectron counting distribution also agrees with the results of Vyas and Singh [8.4] which are obtained analytically.

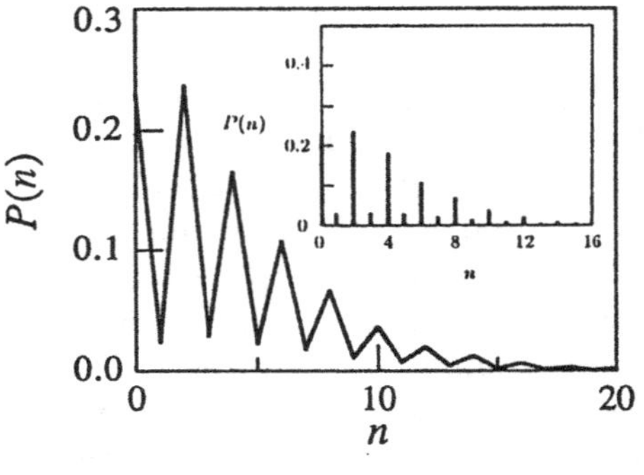

Fig. 8.9. Photoelectron counting distribution for the output of a degenerate parametric oscillator obtained by counting collapses (photon emissions) in the simulation of Fig. 8.8(a). The inset shows the photoelectron counting distribution obtained be other methods [8.3, 8.4].

8.5 Complementary unravellings

In all of the examples we have looked at during this lecture the decomposition of the source master equation dynamics has been based on the direct photoelectric detection of the radiated light. From the stochastic quantum trajectories obtained in this way we can calculate quantities such as average intensities, waiting-time distributions, and photoelectron counting distributions – quantities that are measured by direct photoelectric detection. From the concrete visualization that the quantum trajectory approach allows, we also gain some understanding of the physical processes going on in the source. The decomposition we have used is not, however, unique; it is tailored for direct photoelectric detection. We cannot use the quantum trajectories obtained from this decomposition to calculate everything we might be interested in (at least not in a simple way), nor do these trajectories help us understand every nook and cranny of the quantum dynamics.

In Sect. 7.4 we referred to the decomposition of the source master equation to give quantum trajectories as an *unravelling of the master equation for the source.* The quantum dynamics contained in the master equation are unravelled to give us a picture of what is going on in a visible form. The pictures we have presented so far reveal what is going on when we focus our attention on emitted photons (direct photoelectric detection). Other unravellings of the master equation will give us different pictures, suited to help us understand different aspects of the physics. The complete picture is the complement of all the separate pictures, and by the very nature of quantum mechanics no single picture can substitute for them all. In a way, our difficulty in understanding the full quantum mechanical evolution lies in the fact that the one master equation carries the many pictures forward in parallel. We gain a lot by separating the pictures out.

In the next lecture we will see how to use the quantum trajectory approach to analyze the homodyne detection of squeezed light. By modeling homodyne detection we arrive at a quite different unravelling of the master equation (8.27). In fact, we obtain an infinity of unravellings, one for each choice of the local oscillator phase. As an introduction, Fig. 8.10 shows a sample trajectory for the conditioned mean photon number for two different choices of the local oscillator phase. These correspond to a measurement of the unsqueezed quadrature X and the squeezed quadrature Y of the fluctuating field amplitude. These trajectories look nothing like the trajectory shown in Fig. 8.8(a); they are even qualitatively different from each other, one showing much larger fluctuations than the other. However, all three of these trajectories are equivalent in the mean. They are complementary unravellings of the quantum average $\mathrm{tr}[\rho(t)a^\dagger a]$ (note that it is not the conditioned density operator here); the time average of all three produces exactly the same number.

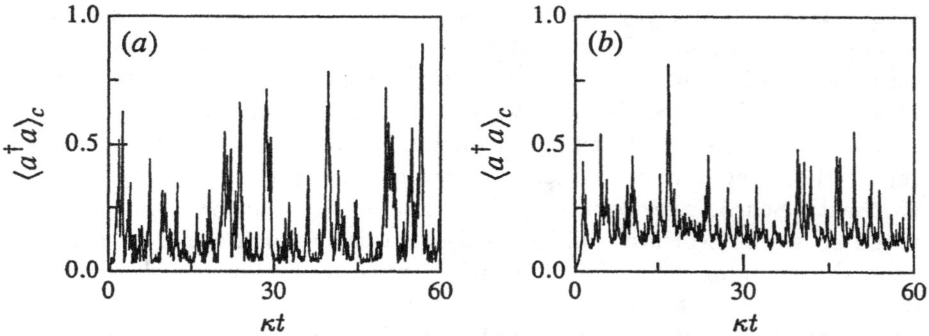

Fig. 8.10. Sample quantum trajectories showing the conditioned mean photon number obtained from the unravelling of the degenerate parametric oscillator master equation described in Sec. 9.2. The parametric oscillator is operated 10% below threshold ($\lambda = 0.9$). (a) The unravelling is based on a measurement of the X-quadrature variance; (b) the unravelling is based on a measurement of the Y-quadrature variance.

References

[8.1] H. J. Carmichael and L. Tian, "Quantum Measurement Theory of Photoelectric Detection," in *OSA Annual Meeting Technical Digest 1990*, Vol. 15 of the OSA Technical Digest Series, Optical Society of America: Washington, D. C., 1990, p. 3.

[8.2] H. J. Carmichael, S. Singh, R. Vyas, and P. R. Rice, *Phys. Rev.* A **39**, 1200 (1989).

[8.3] M. Wolinsky and H. J. Carmichael, "Photoelectron Counting Statistics for the Degenerate Parametric Oscillator," in *Coherence and Quantum Optics VI*, ed. by J. H. Eberly, L. Mandel, and E. Wolf, Plenum: New York, 1989, pp. 1239ff.

[8.4] R. Vyas and S. Singh, *Opt. Lett.* **14**, 1110 (1989); *Phys. Rev.* A **40**, 5147 (1989).

Lecture 9 – Quantum Trajectories III

This lecture is devoted entirely to the degenerate parametric oscillator and the observation of its radiated field by homodyne detection. We hope to accomplish a number of things. First, we will unravel the source master equation [Eq. (8.27)] in a way that is not based on direct photoelectric detection. This provides an explicit example of how different unravellings can be constructed for different measurement schemes to give complementary pictures of a quantized source. Second, we will meet a new method for analyzing quantum trajectories. In this method the stochastic quantum mapping is replaced by a stochastic differential equation for the source wavefunction (a stochastic Schrödinger equation). The method is not always applicable; but when it is, the stochastic differential equation is much easier to simulate than the mapping itself. Third, we will develop a novel approach to the understanding of shot noise reduction in squeezed light measurements. From the point of view of semiclassical photoelectric detection theory, shot noise reduction is a real riddle. We will see how the quantum trajectory approach solves this riddle in a rather simple way, using the collapse of the wavefunction to create nonlocal correlations between the quantum source and the classical photocurrent realized in an observation of the field radiated by the source.

9.1 The riddle of squeezed light

The job of photoelectric detection theory is to set up a relationship between an optical field and a sequence of photoelectron emissions. To the observer the photoelectrons are seen either as a sequence of photoelectric pulses, or as an analogue electric current; these signals are described by classical stochastic process. On the other hand, the optical field that controls the photoelectron emissions is a quantized field. Sometimes, however, we can get away with a description of the optical field in terms of classical stochastics. Then we are using semiclassical photoelectric detection theory. Our first task is to understand why squeezed light, or more specifically, shot noise reduction, is such a riddle from the viewpoint of semiclassical photoelectric detection theory.

In the semiclassical theory of photoelectric detection the emission of photoelectrons is governed by a classical stochastic intensity $\bar{I}(t)$. Throughout this lecture we consider a detector that produces an analogue current, which we describe by a second stochastic process $i(t)$. The theory of photoelectric detection must relate $i(t)$ to $\bar{I}(t)$. The relationship is built up from an understanding of the emission process during short intervals of time Δt. We start by letting Δt be truly infinitesimal in the sense of Sect. 5.1, with a negligible probability for two or more emissions to occur during any Δt. But we can quickly replace this notion with a course-grained dissection of the time. For a detector that produces an analogue current there are many photoelectron emissions during the shortest time interval resolved by the detector (Fig. 6.3). We therefore let Δt be very short compared with the time scale for fluctuations in the optical source, but large enough that many photoelectrons are emitted during Δt. The analysis from Sect. 5.1 now tells us how to construct the current $i(t)$. The instantaneous rate of photoelectron emissions is given by $\xi\bar{I}(t)$, and $\xi\bar{I}(t)\Delta t$ gives the mean number of photoelectrons emitted during the interval $(t, t+\Delta t]$. Then there are fluctuations about this mean. On the scale of Δt the emissions occur randomly. Therefore the fluctuations are Poissonian, and if $\xi\bar{I}(t)\Delta t$ is a large number they are characterized by the Gaussian distribution

$$p(n_t, t, t+\Delta t) = \frac{1}{\sqrt{2\pi\xi\bar{I}(t)\Delta t}} \exp\left(-\frac{1}{2}\frac{\left(n_t - \xi\bar{I}(t)\Delta t\right)^2}{\xi\bar{I}(t)\Delta t}\right). \tag{9.1}$$

Thus, the charge ΔQ emitted from the photocathode during the time Δt is given by

$$\Delta Q/e = \xi\bar{I}(t)\Delta t + \sqrt{\xi\bar{I}(t)}\,\Delta W, \tag{9.2a}$$

where ΔW is a Weiner increment. The photocurrent is now given by

$$i(t)/Ge = \xi\bar{I}(t) + \sqrt{\xi\bar{I}(t)}\,\eta_W(t), \tag{9.2b}$$

where G is a gain factor and $\eta_W(t)$ denotes Gaussian white noise:

$$\overline{\eta_W(t)} = 0, \qquad \overline{\eta_W(t)\eta_W(t')} = \delta(t-t'). \tag{9.3}$$

The Gaussian noise source in (9.3) is the shot noise. Of course, strictly, it does not have an infinite bandwidth. But the white noise idealization is not a limitation for what we are interested in, and it simplifies the mathematics. To incorporate a high-frequency cut-off we would have to model the photocurrent in a more detailed way like we did in Sect. 6.2, and drop the course-grained dissection of time.

The relationship between $\bar{I}(t)$ and $i(t)$ is illustrated schematically in Fig. 9.1. The important observation is that there is additional noise – shot noise – added when $i(t)$ is produced from $\bar{I}(t)$. The photocurrent is not simply a replication of the optical intensity.

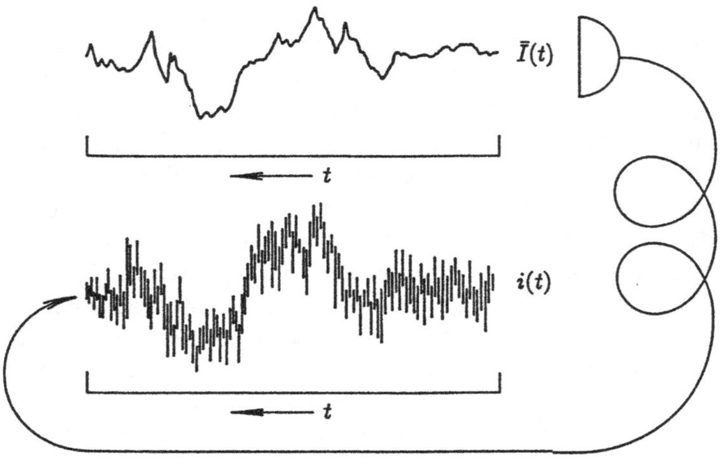

Fig. 9.1. The relationship between an optical intensity $\bar{I}(t)$ and the detected photocurrent $i(t)$. Both are represented as realizations of classical stochastic processes. $\bar{I}(t)$ defines the instantaneous rate function that controls the random emission of photoelectrons that produces $i(t)$.

We now consider homodyne detection (Sect. 6.2). In homodyne detection the photon flux seen by the detector is obtained from the superposition of two fields:

$$\xi\bar{I}(t) = \xi|E_{lo} + e_s(t)|^2, \tag{9.4}$$

where E_{lo} is the constant amplitude of the local oscillator field and $e_s(t)$ is the amplitude of the fluctuating signal field. Under the assumption that $|E_{lo}| \gg |e_s(t)|$, from (9.4) and (9.2b) we have

$$i(t)/Ge = \xi|E_{lo}|^2 + \sqrt{\xi}|E_{lo}|\eta_W(t) + \xi[E_{lo}e_s^*(t) + E_{lo}^*e_s(t)], \tag{9.5}$$

where we have retained the noise terms to dominant order in the amplitude of the local oscillator field. Now $\eta_W(t)$ is Gaussian white noise associated with the random emission of photoelectrons at an average rate that is dominated by the local oscillator photon flux. The signal $e_s(t)$ also introduces noise; this noise has its origin in the source that produces $e_s(t)$. From the viewpoint of semiclassical photoelectric detection theory, the two noise sources have entirely different origins and are surely statistically independent. Then the photocurrent fluctuations $\Delta i(t)/Ge = i(t)/Ge - \xi|E_{lo}|^2$ are characterized by the correlation function

$$\begin{aligned}
&\overline{\Delta i(t)\Delta i(t+\tau)}/(Ge)^2 \\
&\qquad = \xi|E_{lo}|^2\,\overline{\eta_W(t)\eta_W(t+\tau)} + 4\xi^2|E_{lo}|^2\,\overline{e_s^\theta(t)e_s^\theta(t+\tau)} \\
&\qquad = \xi|E_{lo}|^2\delta(\tau) + 4\xi^2|E_{lo}|^2\,\overline{e_s^\theta(t)e_s^\theta(t+\tau)},
\end{aligned} \tag{9.6}$$

where

$$e_s^\theta(t) = \tfrac{1}{2}\left(e_s(t)e^{-i\theta} + e_s^*(t)e^{i\theta}\right), \tag{9.7}$$

and θ is the phase of the local oscillator field. The Fourier transform of (9.6) gives the spectrum of photocurrent fluctuations. The first term on the right-hand side gives the flat shot noise spectrum. The second term must *add noise* to the shot noise level. There is no way that the signal field can bring the total noise below the shot noise level. But this is what happens for squeezed light. Thus, if we retain the picture of photoelectric detection drawn above – random photoelectron emissions over short intervals Δt at an instantaneous rate $\xi\bar{I}(t)$ – how can there ever be shot noise reduction? This is the riddle of squeezed light.

We will solve the riddle during the course of the lecture, but perhaps we can already see what direction to take. The only way in which the above analysis could produce reduced shot noise is if (9.6) is wrong because $\eta_W(t)$ and $e_s(t)$ are correlated. Classical physics provides no mechanism to produce such correlations because $e_s(t)$ is presented, ready made, to the detector, and $\eta_W(t)$ is generated during the detection process itself. But quantum mechanics provides a mechanism. The notion of the collapse of the wavefunction suggests that the emission of each photoelectron at the detector is accompanied by a collapse of the wavefunction that describes the quantum system monitored by the detector. Unpalatable as it is, this collapse must be communicated in a self-consistent way (backwards in time) throughout an extended system, all the way back to the source that produces $e_s(t)$. In this way the quantum state of the source suffers a collapse for every photoelectron emission at the detector. Through the accumulated collapses its radiated field will become correlated with the random fluctuations contained in $\eta_W(t)$. We are going to use the quantum trajectory approach to add quantitative substance to this qualitative picture.

9.2 Homodyne detection

We can use the master equation (8.27) to describe the source of squeezed light. But in place of the decomposition (8.28) and (8.29) we now need a decomposition based on a homodyne detection scheme. This means we must extend our view of the source to include the local oscillator. Figure 9.2 illustrates the model we will use. The model includes two optical cavities: one cavity contains a nonlinear crystal and radiates a beam of squeezed light; the other is prepared in a coherent state and radiates the local oscillator field. The master equation for the complete system is given by

$$\begin{aligned}\dot{\rho} = {}& -i\omega_C[a^\dagger a, \rho] - i\omega_C[b^\dagger b, \rho] + (\kappa\lambda/2)[a^{\dagger 2}e^{-i2\omega_C t} - a^2 e^{i2\omega_C t}, \rho] \\ & + \kappa(2a\rho a^\dagger - a^\dagger a\rho - \rho a^\dagger a) + \gamma(2b\rho b^\dagger - b^\dagger b\rho - \rho b^\dagger b),\end{aligned} \tag{9.8}$$

where $b^\dagger$ and b are creation and annihilation operators for photons in the local oscillator mode, and 2γ is the decay rate for photons in the local oscillator cavity. The initial density operator is

$$\rho(0) = \rho_a \rho_b, \tag{9.9a}$$

with

$$\rho_a = |0\rangle\langle 0|, \qquad \rho_b = |\beta\rangle\langle\beta|; \tag{9.9b}$$

β is the initial amplitude of the local oscillator field. The two output fields are combined by a beam splitter to produce the quantized source field at the detector:

$$\hat{\mathcal{E}}_s = -i\sqrt{R}\sqrt{2\gamma}\, b(t - r/c) + \sqrt{1-R}\sqrt{2\kappa}\, a(t - r/c), \tag{9.10}$$

where R is the reflection coefficient of the beam splitter and we assume that the retardation times from the cavities to the detector are equal.

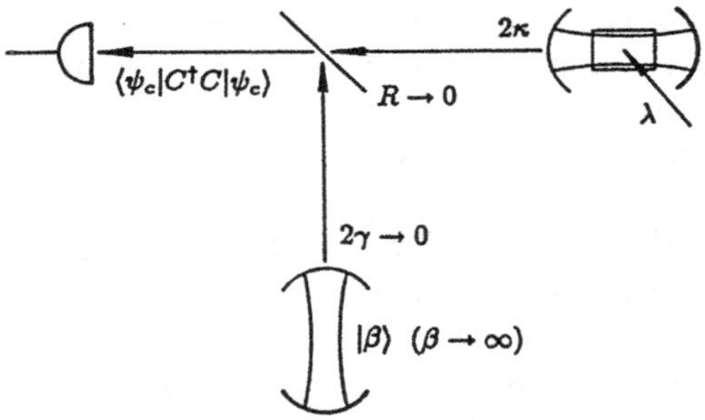

Fig. 9.2. Model of the source seen by the detector in homodyne detection. The pumped cavity is a parametric oscillator, a source of squeezed light. The second cavity radiates a coherent local oscillator field.

We can now decompose the master equation (9.8) along the lines discussed in the previous two lectures. Between collapses the evolution of the unnormalized conditioned density operator $\bar{\rho}_c(t)$ is governed by the superoperator $\mathcal{L} - \mathcal{S}$, where

$$\begin{aligned}(\mathcal{L} - \mathcal{S})\bar{\rho}_c = &-i\omega_C[a^\dagger a, \bar{\rho}_c] - i\omega_C[b^\dagger b, \bar{\rho}_c] \\ &+ (\kappa\lambda/2)[a^{\dagger 2} e^{-i2\omega_C t} - a^2 e^{i2\omega_C t}, \bar{\rho}_c] \\ &+ R(2\kappa) a\bar{\rho}_c a^\dagger - \kappa(a^\dagger a \bar{\rho}_c + \bar{\rho}_c a^\dagger a) \\ &+ (1-R)(2\gamma) b\bar{\rho}_c b^\dagger - \gamma(b^\dagger b\bar{\rho}_c + \bar{\rho}_c b^\dagger b) \\ &+ i\sqrt{R(1-R)}\sqrt{2\gamma}\sqrt{2\kappa}\,(b\bar{\rho}_c a^\dagger - a\bar{\rho}_c b^\dagger).\end{aligned} \tag{9.11}$$

The collapse that accompanies each photoelectron emission is governed by the superoperator $\mathcal{S}$, where

$$\mathcal{S}\bar{\rho}_c = \left(-i\sqrt{R}\sqrt{2\gamma}\, b + \sqrt{1-R}\sqrt{2\kappa}\, a\right)\bar{\rho}_c\left(i\sqrt{R}\sqrt{2\gamma}\, b^\dagger + \sqrt{1-R}\sqrt{2\kappa}\, a^\dagger\right). \tag{9.12}$$

The collapse probability for the interval $(t, t+\Delta t]$ is given by

$$p_c(t) = \mathrm{tr}[\mathcal{S}\rho_c(t)]\Delta t. \tag{9.13}$$

Note that in (9.13), and until it is stated otherwise, Δt is truly infinitesimal in the sense of Sect. 5.1.

As things stand the conditioned density operator does not factorize as a pure state. However, we do not yet have our model in final form. The model has two deficiencies. First, a nonzero reflectivity R for the beam splitter means that some of the squeezed light is lost, which will limit the observed shot noise reduction [9.1, 9.2]. We therefore want to let $R \to 0$; to compensate for this the local oscillator amplitude must become infinite. Second, the amplitude of the initial local oscillator state will decay in time; but we want this amplitude to remain constant throughout the measurements. This is ensured if we let $\gamma \to 0$. This limit also requires the local oscillator amplitude to become infinite. We deal with both deficiencies by taking the limit

$$R \to 0,\ \gamma \to 0,\ \beta \to \infty,\ \text{with } f = R2\gamma|\beta|^2 \text{ constant}; \tag{9.14}$$

f is the local oscillator photon flux at the detector.

In the limit (9.14) the conditioned density operator may be written in the form

$$\rho_c(t) = \left(|e^{-i\omega_C t}\beta\rangle\langle e^{-i\omega_C t}\beta|\right)\rho_c^a(t), \tag{9.15}$$

where $\rho_c^a(t)$ describes the state of the parametric oscillator alone. Now, from (9.11), the evolution of the unnormalized state $\bar{\rho}_c^a(t)$ between collapses is governed by the superoperator $\mathcal{L} - \mathcal{S}$, where

$$\begin{aligned}(\mathcal{L}-\mathcal{S})\bar{\rho}_c^a = &-i\omega_C[a^\dagger a, \bar{\rho}_c] + (\kappa\lambda/2)[a^{\dagger 2}e^{-i2\omega_C t} - a^2 e^{i2\omega_C t}, \bar{\rho}_c^a]\\ &- \kappa(a^\dagger a\bar{\rho}_c^a + \bar{\rho}_c^a a^\dagger a)\\ &- \sqrt{f}\sqrt{2\kappa}(e^{i\theta}e^{-i\omega_C t}\bar{\rho}_c^a a^\dagger + e^{-i\theta}e^{i\omega_C t}a\bar{\rho}_c^a).\end{aligned} \tag{9.16}$$

From (9.12), the collapses are governed by the superoperator $\mathcal{S}$, where

$$\mathcal{S}\bar{\rho}_c^a = \left(\sqrt{f}e^{i\theta}e^{-i\omega_C t} + \sqrt{2\kappa}a\right)\bar{\rho}_c^a\left(\sqrt{f}e^{-i\theta}e^{i\omega_C t} + \sqrt{2\kappa}a^\dagger\right). \tag{9.17}$$

Equations (9.16) and (9.17) allow $\bar{\rho}_c^a(t)$ to be written in terms of a pure state:

$$\bar{\rho}_c^a(t) = |\bar{\psi}_c(t)\rangle\langle\bar{\psi}_c(t)|. \tag{9.18}$$

Then the unnormalized state $|\bar{\psi}_c(t)\rangle$ satisfies the nonunitary Schrödinger equation (8.4a) with non-Hermitian Hamiltonian

$$\begin{aligned}H = &\hbar\omega_C a^\dagger a + i\hbar(\kappa\lambda/2)(a^{\dagger 2}e^{-i2\omega_C t} - a^2 e^{i2\omega_C t})\\ &- i\hbar\kappa a^\dagger a - i\hbar\sqrt{f}e^{-i\theta}e^{i\omega_C t}\sqrt{2\kappa}a.\end{aligned} \tag{9.19}$$

Its evolution is interrupted by collapses

$$|\bar{\psi}_c\rangle \to \hat{C}|\bar{\psi}_c(t)\rangle = \left(\sqrt{f}e^{i\theta}e^{-i\omega_c t} + \sqrt{2\kappa}\,a\right)|\bar{\psi}_c\rangle, \tag{9.20}$$

where the probability for a collapse to occur in the interval $(t, t+\Delta t]$ is given by

$$p_c(t) = \frac{\langle\bar{\psi}_c(t)|(\sqrt{f}e^{-i\theta}e^{i\omega_c t} + \sqrt{2\kappa}\,a^\dagger)(\sqrt{f}e^{i\theta}e^{-i\omega_c t} + \sqrt{2\kappa}\,a)|\bar{\psi}_c(t)\rangle}{\langle\bar{\psi}_c(t)|\bar{\psi}_c(t)\rangle}\Delta t. \tag{9.21}$$

Equations (8.4a) and (9.19)–(9.21) define our unravelling of the source master equation (8.27) based on homodyne detection of the radiated field.

9.3 Nonclassical photoelectron correlations

In a realistic homodyne measurement the local oscillator photon flux f is many orders of magnitude larger than the signal flux $2\kappa\langle\psi_c(t)|a^\dagger a|\psi_c(t)\rangle$. It follows that the change produced in the conditioned state $|\psi_c(t)\rangle$ by the collapse (9.20) is extremely small. Physically this means that a photoelectron emission probably corresponds to the annihilation of a local oscillator photon, with only a small probability, $\sim f/2\kappa\langle\psi_c(t)|a^\dagger a|\psi_c(t)\rangle$, that a photon was annihilated from the signal field; of course, the two possibilities exist as a superposition, not as a classical choice – *either* one *or* the other. Now, although the collapses are very small, on the characteristic time scale $(2\kappa)^{-1}$ for fluctuations in the signal field they occur very often. In the limit $f/2\kappa \to \infty$ the conditioned state $|\psi_c(t)\rangle$ suffers infinitesimal collapses, but at an infinite rate. Clearly this limit is impractical for a numerical simulation that follows every photoelectron emission. We will treat this limit by converting the quantum mapping into a stochastic differential equation. Before we do this, let us look at a few results obtained from the quantum mapping for a less extreme value of $f/2\kappa$.

If we count the photoelectron emissions that occur over a fixed interval T we are effectively integrating the photocurrent $i(t)$. The result of this counting experiment will be different each time we carry it out because $i(t)$ is a stochastic quantity. To dominant order in $f/2\kappa$ the average number of emissions will be fT. In the absence of the squeezed light there will be Poisson fluctuations about this average; the squeezing will change the Poisson distribution. We know that if the phase θ is chosen so that the squeezed quadrature is monitored, the photocurrent noise is reduced below the shot noise level over a bandwidth 2κ about d.c. [Eqs. (6.45) and (6.62b)]. Thus, in this case we expect to obtain a sub-Poisssonian counting distribution when we count photoelectrons for a time longer than the inverse bandwidth of the squeeezing. On the other hand, if the phase θ is chosen to monitor the unsqueezed quadrature, the counting distribution will become super-Poissonian for long counting times.

Results in accord with these expectations are shown in Fig. 9.3. The figure shows the distributions obtained by counting the number of collapses (photoelectron emissions) that occur in each of 10,000 quantum trajectories, for three different counting times. For a Poisson distribution the half-width at half-maximum is given by the square root of the mean; in Fig. 9.3(a) the widths get progressively narrower than this value with increased counting time, while in Fig. 9.3(b) they get progressively broader. It is worthwhile mentioning again just how the narrowing can be achieved. The rate of photoelectron emissions at any instant is determined by (9.21) and is almost equal to f. For times much shorter that $(2\kappa)^{-1}$ these emissions are random. Let us say for arguments sake that over some such interval the number of emissions is much larger than the average number expected. The source knows about this deviation from the norm due to the collapses it has suffered; these collapses adjust the state of the source so that over the longer time scale $\sim (2\kappa)^{-1}$ the interference term in (9.21) is able to bring the number of photoelectron emissions back into line. After we convert the quantum mapping into a stochastic differential equation this communication from the observed photocurrent back to the source will appear explicitly in the equation.

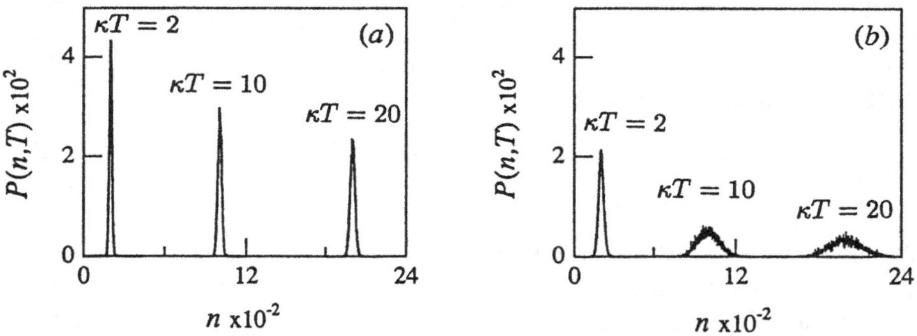

Fig. 9.3. Photoelectron counting distributions for the homodyne detection of squeezed light. (a) Detection of the squeezed Y quadrature ($\theta = \pi/2$). (b) Detection of the unsqueezed X quadrature ($\theta = 0$). The other parameters are $\lambda = 0.5$ and $f/2\kappa = 50$.

9.4 Stochastic Schrödinger equation for the degenerate parametric oscillator

We now shift our viewpoint to match the one which lead us to the semiclassical photocurrent (9.5). We want to take the limit $f/2\kappa \to \infty$. In this limit the conditioned wavefunction suffers an infinite number of infinitesimal collapses in any finite interval $(t, t+\Delta t]$. We will derive a stochastic Schrödinger equation for the conditioned wavefunction for the source, and along with it a quantum-mechanical version of (9.5). The two equations will be coupled; this is a sharp contrast to the semiclassical theory where the definition of the signal field $e_s(t)$ is completely independent of the observed photocurrent $i(t)$.

Our starting point is the quantum mapping for homodyne detection written in the form (7.32). The calculation is simpler, however, if we leave out the normalization of the state and replace it explicitly at the end. We therefore start from the following mapping. If t_n and t_{n+1} are the times of two successive collapses – $t_n + r/c$ and $t_{n+1} + r/c$ are the times of two successive photoelectron emissions – and $|\bar{\psi}_c(t_n)\rangle$ and $|\bar{\psi}_c(t_{n+1})\rangle$ are the unnormalized conditioned wavefunctions immediately after these collapses, then

$$|\bar{\psi}_c(t_{n+1})\rangle = \hat{C}e^{-(i/\hbar)H\tau_{n+1}}|\bar{\psi}_c(t_n)\rangle, \tag{9.22}$$

where $\tau_{n+1} = t_{n+1} - t_n$ is a random time. In the present example, from (9.19) and (9.20) we have

$$H = i\hbar(\kappa\lambda/2)(a^{\dagger 2} - a^2) - i\hbar a^\dagger a - i\hbar\sqrt{f}e^{-i\theta}\sqrt{2\kappa}\,a, \tag{9.23a}$$
$$\hat{C} = \sqrt{f}e^{i\theta} + \sqrt{2\kappa}\,a. \tag{9.23b}$$

We have transformed to the interaction picture so that these operators are no longer explicitly dependent on time.

Now for $f/2\kappa \gg 1$ the conditioned wavefunction only accumulates a significant change after very many iterations of the mapping (9.22). We therefore consider

$$\Delta|\bar{\psi}_c\rangle = |\bar{\psi}_c(t_{n+m})\rangle - |\bar{\psi}_c(t_n)\rangle, \tag{9.24}$$

where m is a large (random) number defined by the requirement

$$\tau_{n+1} + \tau_{n+2} + \cdots + \tau_{n+m} = \Delta t. \tag{9.25}$$

$\Delta|\bar{\psi}_c\rangle$ is the change in the conditioned wavefunction during the interval $(t, t+\Delta t] \equiv (t_n, t_{n+m}]$. Following the discussion above (9.1) we assume that Δt is short compared to the time scale for significant change to occur in the state of the source, but long compared to the average time between collapses (photoelectron emissions). Clearly, m is the number of collapses that occur in the interval $(t, t + \Delta t]$, or, equivalently, the number of photoelectron emissions in the interval $(t + r/c, t + \Delta t + r/c]$. Since Δt is an intermediate time scale, there are very many collapses during Δt, and the collapses occur randomly in time at a rate $\langle\psi_c(t)|\hat{C}^\dagger\hat{C}|\psi_c(t)\rangle$. Thus, corresponding to the semiclassical result (9.1), m is to be chosen from the Gaussian distribution

$$p(m,t,t+\Delta t) = \frac{1}{\sqrt{2\pi\langle(\hat{C}^\dagger\hat{C})(t)\rangle_c\Delta t}} \exp\left[-\frac{1}{2}\frac{\left(m - \langle(\hat{C}^\dagger\hat{C})(t)\rangle_c\Delta t\right)^2}{\langle(\hat{C}^\dagger\hat{C})(t)\rangle_c\Delta t}\right], \tag{9.26a}$$

where

$$\langle(\hat{C}^\dagger\hat{C})(t)\rangle_c \equiv \langle\psi_c(t)|\hat{C}^\dagger\hat{C}|\psi_c(t)\rangle. \tag{9.26b}$$

In the language of stochastic processes we write

$$m = \langle(\hat{C}^\dagger\hat{C})(t)\rangle_c\Delta t + \sqrt{\langle(\hat{C}^\dagger\hat{C})(t)\rangle_c}\,\Delta W, \tag{9.27}$$

which is the quantum-mechanical replacement for (9.2a); ΔW is a Weiner increment.

Our stochastic Schrödinger equation is derived from (9.24). We do not have time for the details of the calculation and therefore just note the main steps: (i) We expand (9.22) for small $\tau_{n+1} \sim 1/f$. This gives an expansion in powers of $\sqrt{2\kappa/f}$ in which we keep terms of order unity, $\sqrt{2\kappa/f}$, and $2\kappa/f$. (ii) We then calculate $\Delta|\bar{\psi}_c(t)\rangle$ from (9.24) by iterating the expanded mapping and keeping terms to the same order as before. After this step $\Delta|\bar{\psi}_c(t)\rangle$ depends explicitly on m. (iii) We substitute (9.27) for m with $\hat{C}$ substituted from (9.23b) and take the limit $f/2\kappa \to \infty$. (iv) We finally let $\Delta t \to dt$, $\Delta W \to dW$ and $(\Delta W)^2 \to dt$. The result is a stochastic differential equation for the unnormalized conditioned state of the source:

$$\frac{d}{dt}|\bar{\psi}_c\rangle = \frac{1}{i\hbar}H_W(t)|\bar{\psi}_c\rangle, \tag{9.28}$$

where $H_W(t)$ is the stochastic, non-Hermitian Hamiltonian

$$\begin{aligned} H_W(t) = {}& i\hbar(\kappa\lambda/2)(a^{\dagger 2} - a^2) - i\hbar\kappa a^\dagger a \\ & + i\hbar\left[\sqrt{2\kappa}\,\langle(e^{i\theta}a^\dagger + e^{-i\theta}a)(t)\rangle_c + \eta_W(t + r/c)\right]e^{-i\theta}\sqrt{2\kappa}\,a, \end{aligned} \tag{9.29a}$$

where

$$\langle(e^{i\theta}a^\dagger + e^{-i\theta}a)(t)\rangle_c \equiv \langle\psi_c(t)|(e^{i\theta}a^\dagger + e^{-i\theta}a)|\psi_c(t)\rangle \tag{9.29b}$$

and $\eta_W(t + r/c)$ is a Gaussian white noise.

Equation (9.27) gives the photocurrent. Recall that m is the number of photoelectrons emitted in the interval $(t + r/c, t + \Delta t + r/c]$. Therefore, after substituting for $\hat{C}$ and keeping terms proportional to f and $\sqrt{f}$, the observed photocurrent is

$$i(t+r/c)/Ge = f + \sqrt{f}\left[\eta_W(t+r/c) + \sqrt{2\kappa}\,\langle(e^{i\theta}a^\dagger + e^{-i\theta}a)(t)\rangle_c\right]. \quad (9.30)$$

This is the quantum mechanical version of (9.5). It is precisely the same expression with the substitutions $\sqrt{\xi}\,E_{lo} \to \sqrt{f}e^{i\theta}$ and $\sqrt{\xi}\,e_s(t) \to \sqrt{2\kappa}\,\langle a(t-r/c)\rangle_c = \sqrt{2\kappa}\,\langle\psi_c(t-r/c)|a|\psi_c(t-r/c)\rangle$. We use the argument $t+r/c$ for the white noise η_W to remind ourselves that this process entered to describe the randomness of photoelectron emissions at the detector. Mathematically, the important point regarding this noise source is that it appears in both (9.29b) and (9.30), *evaluated at the same time*. So far as the mathematics is concerned, the argument of η_W could just as well be t. We will say more about this shortly.

We can now see that the picture of homodyne detection obtained from the quantum trajectory approach is essentially the same as the one obtained in Sect. 9.1 from the semiclassical theory of photoelectric detection. The photocurrent is produced by random photoelectron emissions over short intervals Δt at a rate determined by the instantaneous photon flux illuminating the detector. From the randomness of the photoelectron emissions the photocurrent $i(t+r/c)$ acquires a noise component $\eta_W(t+r/c)$. The only difference between the quantum and semiclassical theories is that, in the quantum trajectory theory, the photon flux [Eq. (9.26b)] depends on a conditioned wavefunction that satisfies the Schrödinger equation (9.28). This Schrödinger equation incorporates the effects of the wavefunction collapses that accompany photoelectron emission, and therefore depends explicitly on the noise source $\eta_W(t+r/c)$. As a result, the two noise sources that appear in the expression for the photocurrent become correlated.

It is straightforward to use (9.28)–(9.30) to simulate the observed photocurrent. The simulations can be used to compute correlation functions and spectra for the photocurrent noise – $\Delta i(t)/Ge\sqrt{f} = (1/\sqrt{f})[i(t)/Ge - f]$ – as if they were signals measured in an experiment. Figures 9.4 and 9.5 show results obtained in this way for the squeezed and unsqueezed quadratures of the field radiated by a degenerate parametric oscillator below threshold. Figure 9.6 shows examples of the fluctuating conditioned field amplitudes $\langle(e^{i\theta}a^\dagger + e^{-i\theta}a)(t)\rangle_c$. These emphasize again the complementary nature of the pictures obtained from different unravellings of a source master equation (Sect. 8.5). In contrast to Fig. 9.6, the conditioned field amplitude is zero at all times for the unravelling based on direct photoelectric detection.

These computations, and the above theory, assume perfect detection efficiency. It is not difficult to generalize the method for an imperfect detector. All that happens is that the noise $\eta_W(t+r/c)$ in (9.29a) is replaced by two uncorrelated noise sources added in the proportion η_d and $1-\eta_d$, where η_d is the detector efficiency. One of these is the noise source that appears in the photocurrent, the other is not (it describes unobserved collapses). It

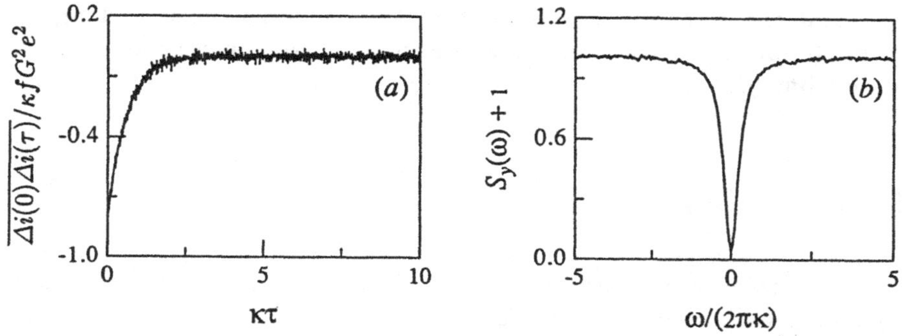

Fig. 9.4. (a) Photocurrent correlation function and (b) spectrum of photocurrent fluctuations for the homodyne detection of the squeezed output of a degenerate parametric oscillator operated 30% below threshold ($\lambda = 0.7$). The squeezed Y quadrature is measured ($\theta = \pi/2$).

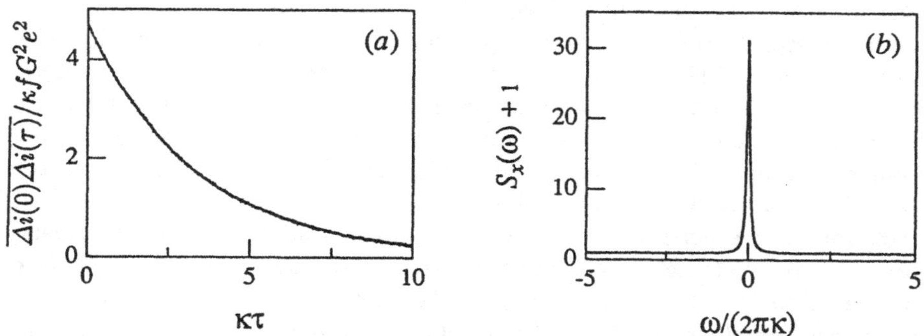

Fig. 9.5. (a) Photocurrent correlation function and (b) spectrum of photocurrent fluctuations for the homodyne detection of the squeezed output of a degenerate parametric oscillator operated 30% below threshold ($\lambda = 0.7$). The unsqueezed X quadrature is measured ($\theta = 0$).

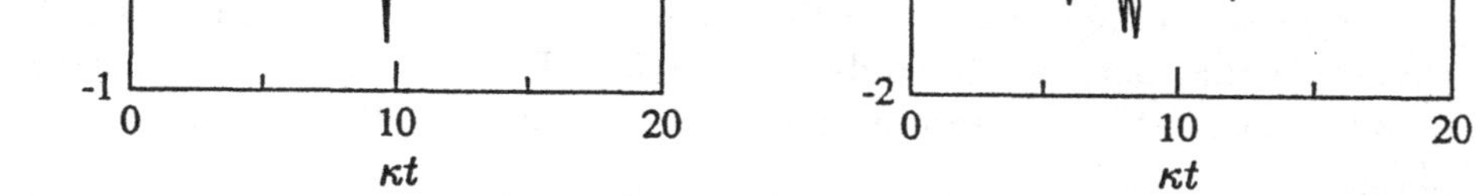

Fig. 9.6. Sample quantum trajectories generated by (9.28)–(9.30) showing the conditioned mean field quadrature amplitudes for a degenerate parametric oscillator operated 70% below threshold ($\lambda = 0.7$). (a) The Y amplitude for Y-quadrature homodyne detection ($\theta = \pi/2$). (b) The X amplitude for X-quadrature homodyne detection ($\theta = 0$).

follows that the correlations between the Gaussian white noise and signal noise in the photocurrent are less strong, and the shot noise reduction is correspondingly less.

9.5 Nonlocality

We conclude this lecture with some observations about the general structure of the theory we have developed. We stated in Sect. 9.1 that it is the purpose of photoelectric detection theory to relate an optical field to a sequence of photoelectron emissions. In the case of semiclassical photoelectric detection theory the relationship is one between two classical stochastic processes. In the full quantum mechanical theory it is a relationship between a classical stochastic process and a quantized field.

In the standard formulation of photoelectric detection theory the relationship is established at the level of correlation functions; correlation functions for the classical photocurrent are related to correlation functions for the quantized field. Using the quantum trajectory approach we get something that goes a little deeper. We essentially set up an interface at the level of equations of motion – an interface between a wavefunction evolving according to a stochastic Schrödinger equation, and a classical stochastic photocurrent. Setting up an interface like this is always a little awkward because of the fundamental incompatibility between the mathematical language used on its two sides. The neoclassical theory of radiative interactions illustrates the difficulty quite well [9.3]. This theory couples quantized matter equations to the classical Maxwell's equations by using the *mean* polarization of the material as a source in Maxwell's equations. The theory is only partially successful; one obvious deficiency is that it does not transfer the fluctuations of the quantized sources to the field. Photoelectric detection goes in the reverse direction; the interface is between a quantized field equation and a classical description for the matter (electric current). The idea, however, is similar, and in contrast to neoclassical theory, the quantum trajectory approach to photoelectric detection rigorously transfers the quantum fluctuations to the classical current. It does this by using a *stochastic* conditioned average to coupled the quantum mechanical equations to the classical equations. For homodyne detection the stochastic average is the quantity inside the square brackets in (9.29a), and (9.30) provides the coupling.

Just how far can we extend the classical ideas in this theory? We have not replaced quantum mechanics by a classical stochastic process; we have simply formulated our description of the quantum mechanical world in such a way that it has (stochastic) classical appendages that a classical world can recognize and hold on to. Of course, we might choose to view the appendages as the only known reality, and relegate the quantum mechanical body to which they are attached to some unknown and impenetrable world; with respect to Eqs. (9.28)-(9.30), we might regard the Schrödinger equation (9.28) as nothing more that an elaborate algorithm for advancing the classical quantity $\langle (e^{i\theta}a^\dagger + e^{-i\theta}a)(t)\rangle_c$ in time. With this view we do, in fact, replace quantum mechanics by a classical stochastic process (actually many complementary processes). If we adopt this viewpoint, do all the peculiarities of quantum mechanics disappear? They do not. We must still accept, or somehow circumvent, a manifest nonlocality in time.

This nonlocality becomes very clear if we use (9.30) to write the stochastic, non-Hermitian Hamiltonian (9.29a) in the form

$$H_W(t) = i\hbar(\kappa\lambda/2)(a^{\dagger 2} - a^2) - i\hbar\kappa a^\dagger a - i\hbar\frac{i(t+r/c)/Ge - f}{\sqrt{f}}e^{\perp i\theta}\sqrt{2\kappa}\,a. \tag{9.31}$$

We see here that the evolution of the source does not occur independently of the observed photocurrent. Most importantly, the source *anticipates* the noise that will be observed in the photocurrent a time r/c in the future. This would be fine if we could say that the photocurrent fluctuations are simply a transcription of the field fluctuations produced by the source; the advanced time argument on the photocurrent in (9.31) is then a trivial consequence of the transformation between a field located at the source and the same field located a time r/c later at the detector. But we have not viewed the photocurrent fluctuations as a direct transcription of the field fluctuations. The $\eta_W(t+r/c)$ component of the fluctuations in (9.30) came from the randomness of photoelectron emissions at the detector, communicated backwards in time to the source by the collapse of the wavefunction. Thus, we preserve the semiclassical view of random photoelectron emissions at a rate determined by the instantaneous intensity (now a conditioned quantum average) at the expense of introducing a nonlocality in time.

We can circumvent this problem by regarding the Gaussian white noise (the collapses) to originate at the source. We would then replace $\eta_W(t+r/c)$ by $\eta_W(t)$ in both (9.29a) and (9.30). But now there is a new problem; now the field illuminating the detector must explicitly orchestrate the times of the photoelectron emissions so that the $\eta_W(t)$ in the photocurrent $i(t+r/c)$ is a precise transcription of the $\eta_W(t)$ generated by collapses at the source. The conventional formulation of quantum mechanics does not provide a mechanism for doing this. Perhaps it can be done in the quantum trajectory formulation. For example, each collapse of the source wavefunction introduces a small discontinuity into the conditioned photon flux [Eq. (9.26b)]. This discontinuity could signal photon arrival times to the detector, telling the detector when photoelectron emissions must occur. This would not work for a coherent source since a coherent state collapses to itself, and there are no discontinuities. But a variation on the idea could be concocted to cover the coherent source case. We will not pursue such inventions here. It is worthwhile raising these issues, however, to show that the interpretational difficulties we have come to expect from quantum mechanics are still there, just below the surface. Actually, it is a pleasing feature of the quantum trajectory approach that an equation like (9.31) states these difficulties in such a clear manner.

References

[9.1] M. J. Collett and C. W. Gardiner, *Phys. Rev.* A **30**, 1386 (1984).
[9.2] J. H. Shapiro, H. P. Yuen, and J. A. Machado Mata, *IEEE Trans. Inf. Th.*, Vol **IT-25**, 179 (1979).
[9.3] C. R. Stroud, Jr. and E. T. Jaynes, *Phys. Rev.* A **1**, 106 (1970).

Lecture 10 – Quantum Trajectories IV

In this final lecture we are going to talk about applications of the quantum trajectory approach. To be more precise, we will talk about one area of current research where the standard methods of analysis discussed in Lectures 3 and 4 are either invalid or difficult to apply, and where the quantum trajectory approach provides a new, and perhaps very useful way to proceed. The area of research is cavity quantum electrodynamics (cavity Q.E.D.). The physical system we consider is an optical cavity containing a single two-state atom, driven by a coherent field resonant with the atom and one mode of the cavity. If the interaction between the atom and the cavity mode is treated semiclassically, the presence of the atom is accounted for by a nonlinear susceptibility; in this approximation the system exhibits absorptive optical bistability. The first step beyond the semiclassical approximation introduces quantum fluctuations in the manner described in Sects. 4.4 and 4.5, where a small Gaussian "fuzz-ball" smears out the semiclassically determined states. We will be interested in situations where the "fuzz-ball" becomes very large compared with the scale of the semiclassical nonlinear physics. In these situations the approximations that give rise to the "fuzz-ball" picture break down. The quantum trajectory approach provides a picture of the quantum fluctuations that is not limited in this way.

10.1 Single-atom absorptive optical bistability

Let us begin with a brief review of the semiclassical theory of optical bistability for a two-state medium [10.1, 10.2]. Consider a collection of N atoms distributed uniformly throughout an interaction volume V inside an optical cavity. The atoms have a resonance frequency ω_A and they interact with one mode of the cavity with resonance frequency ω_C. The cavity is illuminated by a coherent field of frequency ω. Since the atoms respond in a nonlinear way to the field that drives them, the strength of the field inside the cavity, and hence, the strength of the field transmitted by the cavity, must be determined in a self-consistent way. Assume that the field inside the cavity is $\langle a(t)\rangle$ – the time dependence includes the harmonic oscillation at frequency ω and the field amplitude is measured in photon number units. Then, in steady state, the single-atom polarization is

$$\langle\sigma_-(t)\rangle = -\frac{2g}{\gamma}\frac{1-i\delta}{1+\delta^2+n_{sat}^{-1}|\langle a\rangle|^2}\langle a(t)\rangle, \tag{10.1}$$

where $\gamma/2$ is the atomic linewidth (half-width at half-maximum), $\delta = 2(\omega_A - \omega)/\gamma$,

$$g = \left(\frac{\omega|\mu|^2}{2\hbar\epsilon_0 V}\right)^{1/2} \tag{10.2}$$

is the dipole coupling constant, where μ is atomic dipole moment, and

$$n_{sat} = \gamma^2/8g^2 \tag{10.3}$$

is the *saturation photon number*. The polarization (10.1) radiates into the cavity mode so that the steady-state field inside the cavity is given by

$$\langle a(t)\rangle = \frac{\mathcal{E}e^{-i\omega t} + Ng\langle\sigma_-(t)\rangle}{\kappa(1+i\phi)}, \tag{10.4}$$

where κ is the cavity linewidth, $\phi = (\omega_C - \omega)/\kappa$, and $\mathcal{E}$ is the amplitude of the driving field. [$(\mathcal{E}/\kappa)^2$ is the number of photons inside the cavity in steady state when the atoms are removed.] The requirement that (10.1) and (10.4) both be true gives the optical bistability state equation

$$\begin{aligned} &n_{sat}^{-1}(\mathcal{E}/\kappa)^2 \\ &= n_{sat}^{-1}|\langle a\rangle|^2\left[\left(1+\frac{2C}{1+\delta^2+n_{sat}^{-1}|\langle a\rangle|^2}\right)^2 + \left(\phi - \frac{2C\delta}{1+\delta^2+n_{sat}^{-1}|\langle a\rangle|^2}\right)^2\right], \end{aligned} \tag{10.5}$$

where

$$C = Ng^2/\gamma\kappa \tag{10.6}$$

is the so-called cooperativity parameter.

In the semiclassical approximation (10.5) holds for one atom or many atoms alike. But, actually, as the number of atoms decreases the validity of the semiclassical approximation becomes suspect; the system, in some sense, becomes smaller, and fluctuations should then become more important. To treat the fluctuations we need a microscopic model. For one atom, and for exact resonance ($\delta = \phi = 0$), the microscopic model for optical bistability is provided by the source master equation

$$\begin{aligned} \dot\rho = &-i\tfrac{1}{2}\omega_C[\sigma_z,\rho] - i\omega_C[a^\dagger a,\rho] + g[a^\dagger\sigma_+ - a\sigma_-,\rho] \\ &+ \mathcal{E}[a^\dagger e^{-i\omega_C t} - ae^{i\omega_C t},\rho] \\ &+ (\gamma/2)(2\sigma_-\rho\sigma_+ - \sigma_+\sigma_-\rho - \rho\sigma_+\sigma_-) + \kappa(2a\rho a^\dagger - a^\dagger a\rho - \rho a^\dagger a). \end{aligned} \tag{10.7}$$

This source radiates three fields: The cavity radiates transmitted and reflected fields which are calculated as in Sect. 1.4 using appropriate decay rates $2\kappa_t$ and $2\kappa_r$ for each mirror ($2\kappa_t + 2\kappa_r = 2\kappa$). The third field is radiated out the sides of the cavity by the atom, and is given by (2.61) (we assume the cavity mode subtends a negligible solid angle). Equation (10.7) is the starting point for the calculations discussed in this lecture.

The standard analysis based on the quantum-classical correspondence (Lecture 4) was applied extensively to optical bistability in the 1980s [10.2, 10.3]. This analysis is not applicable here. The reason for this is that, for the atomic variables at least, we cannot identify a scaling parameter to justify a system size expansion (4.4). Compounding this problem is the knowledge that the quantum fluctuations are nonclassical; it is known that optical bistability produces photon antibunching [10.4] and squeezing [10.5]. It follows that the fluctuations do not really fit the classical mold that motivates the quantum-classical correspondence. In particular, when the quantum fluctuations are large something like the positive P representation [10.6] is needed to accommodate the nonclassical noise [10.7]. But this representation has its own difficulties [10.7-10.9]. What is needed then is a direct solution to the operator master equation, or a stochastic formulation based on a true quantum dynamic rather than analogies with classical statistics – the quantum trajectory approach.

The solution to (10.7) can be obtained numerically. However, this easily becomes a very large numerical problem. If n_{max} is the largest photon number kept in a truncated Fock state basis, there are $(2n_{max}+1)(n_{max}+2)$ independent matrix elements in the representation of ρ. Two hundred photon states gives us a system of 10^5 coupled equations. On the other hand, the quantum trajectory approach requires only 400 equations for the same 200 Fock states because it can be formulated in terms of a wavefunction instead of a density matrix. Of course, there is a down side, since long simulations are needed to compute time averages. Nevertheless, the quantum trajectory approach clearly has computational potential that should be explored. Work in this direction is just beginning, therefore the results which follow are only indicative of what can be done and no conclusions will be drawn.

Savage and Carmichael solved (10.7) numerically in a standard way for parameters where the "fuzz-ball" begins to be large on the scale of the nonlinear physics [10.10]. Figure 10.1 shows two Q functions obtained by these authors. The Q functions are bimodal with maxima located in the vicinity of the steady states given by the semiclassical equation (10.5).

To provide a simulation based on the quantum trajectory approach we divide the evolution of the conditioned density operator up into an evolution between collapses, governed by the superoperator $\mathcal{L} - \mathcal{S}_A - \mathcal{S}_C$, where

$$\begin{aligned}(\mathcal{L} - \mathcal{S}_A - \mathcal{S}_C)\bar{\rho}_c = &-i\tfrac{1}{2}\omega_C[\sigma_z, \bar{\rho}_c] - i\omega_C[a^\dagger a, \bar{\rho}_c] + g[a^\dagger\sigma_+ - a\sigma_-, \bar{\rho}_c] \\ &+ \mathcal{E}[a^\dagger e^{-i\omega_C t} - ae^{i\omega_C t}, \bar{\rho}_c] \\ &- (\gamma/2)(\sigma_+\sigma_-\bar{\rho}_c + \bar{\rho}_c\sigma_+\sigma_-) - \kappa(a^\dagger a\bar{\rho}_c + \bar{\rho}_c a^\dagger a),\end{aligned} \tag{10.8}$$

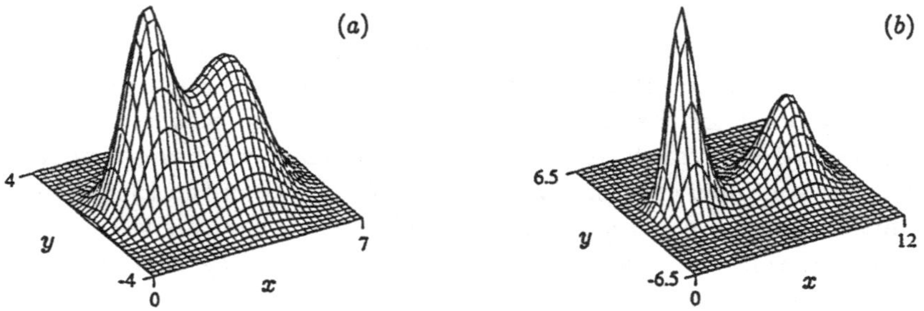

Fig. 10.1. Q functions for single-atom absorptive optical bistability with $C = 6$ and (a) $n_s = 1$, $n_s^{-1/2}(\mathcal{E}/\kappa) = 7.2$; (b) $n_s = 5$, $n_s^{-1/2}(\mathcal{E}/\kappa) = 6.85$. x and y are the real and imaginary parts of the complex field amplitude.

and two types of collapse: for photons that leave through the cavity mirrors we have the collapse operator $\mathcal{S}_C$, where

$$\mathcal{S}_C\bar{\rho}_c = 2\kappa a\bar{\rho}_c a^\dagger, \tag{10.9a}$$

while for photons that leave as fluorescence out the sides of the cavity we have the collapse operator $\mathcal{S}_A$, where

$$\mathcal{S}_A\bar{\rho}_c = \gamma\sigma_-\bar{\rho}_c\sigma_+. \tag{10.9b}$$

For the factorized conditioned density operator the unnormalized conditioned wavefunction obeys the nonunitary Schrödinger equation (8.4a), with non-Hermitian Hamiltonian

$$\begin{aligned} H = {} & \tfrac{1}{2}\hbar\omega_C\sigma_z + \hbar\omega_C a^\dagger a + i\hbar g(a\sigma_+ - a^\dagger\sigma_-) + i\hbar\mathcal{E}(ae^{i\omega_C t} - a^\dagger e^{-i\omega_C t}) \\ & - i\hbar\frac{\gamma}{2}\sigma_+\sigma_- - i\hbar\kappa a^\dagger a. \end{aligned} \tag{10.10}$$

We compute two collapse probabilities for each time step Δt:

$$p_c^C(t) = (2\kappa\Delta t)\frac{\langle\bar{\psi}_c(t)|a^\dagger a|\bar{\psi}_c(t)\rangle}{\langle\bar{\psi}_c(t)|\bar{\psi}_c(t)\rangle}, \tag{10.11a}$$

$$p_c^A(t) = (\gamma\Delta t)\frac{\langle\bar{\psi}_c(t)|\sigma_+\sigma_-|\bar{\psi}_c(t)\rangle}{\langle\bar{\psi}_c(t)|\bar{\psi}_c(t)\rangle}. \tag{10.11b}$$

The corresponding collapse operations are

$$|\bar{\psi}_c\rangle \to \sqrt{2\kappa}a|\bar{\psi}_c\rangle, \tag{10.12a}$$

$$|\bar{\psi}_c\rangle \to \sqrt{\gamma}\sigma_-|\bar{\psi}_c\rangle. \tag{10.12b}$$

This unravelling of the master equation (10.7) is based on direct photoelectric detection. We should note that other unravellings are possible, for example, one based on the homodyne detection scheme discussed in Lecture 9. We will mention a third example later in the lecture.

Results obtained for single-atom optical bistability using the quantum trajectory approach are illustrated in Fig. 10.2. Figure 10.2(a) shows a short section of a time series for the conditioned mean photon number. The values of C and n_{sat} are the same as in Fig. 10.1(a). The distinction between a low intensity state and a high intensity state is clearly visible; but the fluctuations are very large, particularly in the high intensity state. In Figure 10.2(b) a histogram (probability distribution) for the conditioned intensity is constructed from a single time series. Here the two states are very clearly defined. It would be an interesting exercise to compare distributions obtained in this way with those obtained using the standard approximation schemes based on the quantum-classical correspondence. At the moment virtually nothing is known about the relationship between quantum trajectories and the stochastic differential equations obtained using the quantum classical correspondence.

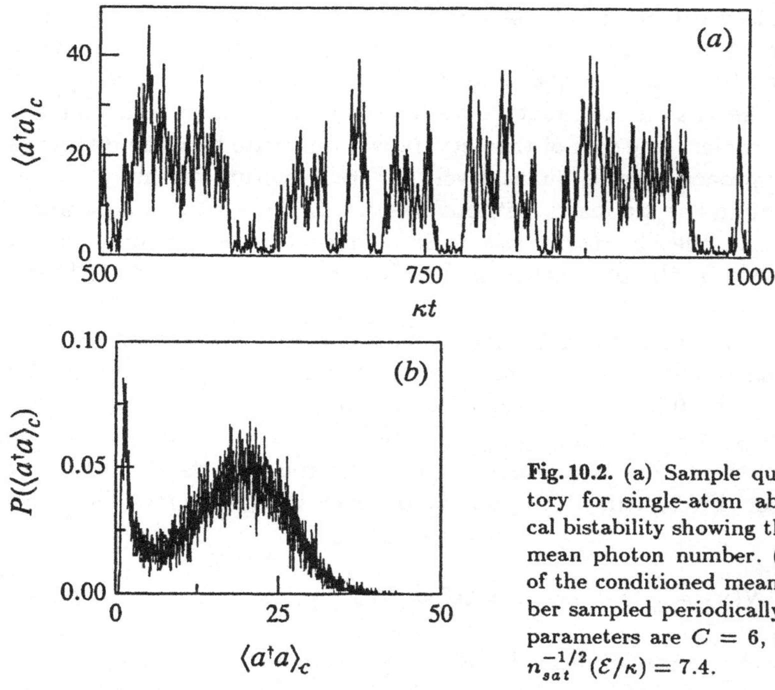

Fig. 10.2. (a) Sample quantum trajectory for single-atom absorptive optical bistability showing the conditioned mean photon number. (b) Histogram of the conditioned mean photon number sampled periodically in time. The parameters are $C = 6$, $n_{sat} = 1$, and $n_{sat}^{-1/2}(\mathcal{E}/\kappa) = 7.4$.

10.2 Strong coupling: cavity Q.E.D.

Aside from the computational advantage of working with wavefunctions rather than density matrices, the quantum trajectory approach has a more fundamental contribution to make. We mentioned above that the quantum fluctuations in optical bistability are nonclassical. This means that the Glauber-Sudarshan representation (Sect. 4.1) does not transform the source master equation into an acceptable Fokker-Planck equation. When the system size expansion is valid this is not necessarily a difficulty, because use of the Q representation, the Wigner representation, or the positive P representation can solve the problem. But when the system size expansion is not valid, none of these representations is guaranteed to give an acceptable stochastic formulation of the quantum statistics. Basically, it seems that there is a level at which the quantum fluctuations must assert their uniquely *quantum* character. Then they are not easily forced into a classical mold; the Fokker-Planck model sets too rigid a constraint on the form of the quantum dynamics. In contrast, the quantum trajectory approach is built from the beginning on quantum mechanical ideas. It is therefore able to provide a stochastic formulation without imposing constraints on the quantum dynamics. The rest of this lecture will illustrate how the quantum trajectory approach gives a qualitatively different picture of the quantum fluctuations than the standard methods based on Fokker-Planck equations. Before we begin the illustration we make a short diversion to understand a little more about the physical regime where the standard methods break down.

What we have to say can be stated with reference to optical bistability; but perhaps a laser model will be more familiar. Consider the model illustrated in Fig. 10.3. Here N atoms interact with a single laser mode containing n photons; γ_p is a pumping rate, and g, κ and $\gamma/2$ have the same meanings as before. Now two principle conditions must be met to construct a normal laser. First, it must be possible to reach the laser threshold. This requires

$$n(4g^2/\gamma)N(p_+ - p_-) = 2\kappa n \Longrightarrow 2C = 2Ng^2/\gamma\kappa \sim 1; \tag{10.13}$$

p_+ and p_- are the probabilities for an atom to be in the upper and lower lasing levels, respectively. Equation (10.13) simply equates the difference between the stimulated emission and absorption rtes to the cavity loss rate. There is then a second, implicit, requirement. The idea with a laser is to achieve "Light Amplification by *Stimulated Emission* of Radiation." If stimulated emission is to dominate spontaneous emission the laser transition must remain unsaturated in the presence of many photons; certainly this is required if the laser is to radiate a large photon flux. Thus, we need

$$n_{sat} = \gamma(\gamma + \gamma_p)/16g^2 \gg 1. \tag{10.14}$$

If, for simplicity, we now take $\gamma_{\wp} \sim \gamma \sim 2\kappa$, (10.13) and (10.14) tell us that a normal laser operates under conditions of weak dipole coupling using very many atoms:

$$g \ll \kappa, \gamma/2, \qquad N \gg 1. \tag{10.15}$$

These are the conditions that produce small quantum noise and justify the system size expansion. Equation (10.14) states that many photons are required to probe the nonlinearity that sets the stable laser operating condition. Taken with (10.13) it leads to the conclusion that many atoms are needed to produce the many photons. Thus, a conventional laser is inherently a many particle device. The average, macroscopic behavior of the device is built up from many single particle contributions. The quantum fluctuations are what remains of the underlying single particle behavior – they evidence the microscopic graininess caused by one photon coming or going, or one atom making a transition. Since one photon or one atom is of little consequence against the background of many particles the fluctuations are small.

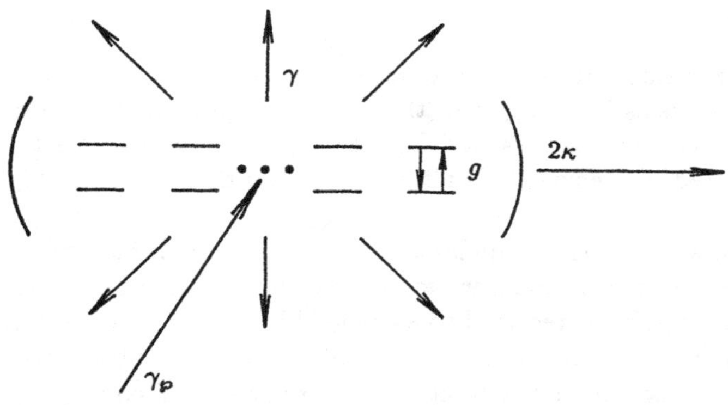

Fig. 10.3. Single-mode laser model. The parameters are defined in the text.

From (10.15) we see that changing conditions (10.13) and (10.14) is ultimately a requirement for strong rather than weak coupling. If we have $2g/\gamma > 1$ and $g/\kappa > 1$ the saturation photon number is small, and one, or even less than one (on average), photon will begin to saturate an atom. It also follows that for just one atom

$$C = C_1 \equiv g^2/\gamma\kappa \tag{10.16}$$

is large, and therefore what the one atom does significantly affects the field to which it couples. This is the regime of *cavity Q.E.D.*.

We have already entered this regime to some extent with the results shown in Figs. 10.1 and 10.2. The values of n_{sat} and $C = C_1$ used there give $g/\kappa = 6 \times \sqrt{8}$ (Fig. 10.1) and $g/\kappa = 6 \times \sqrt{40}$ (Fig. 10.2); although,

$2g/\gamma$ is still less than unity ($2g/\gamma = 1/\sqrt{2} and 1/\sqrt{10}$ respectively). Work in cavity Q.E.D. has primarily been concerned with two parameter regimes: $\kappa \gg g \gg \gamma/2$, which is the parameter regime of cavity-enhanced and -inhibited spontaneous emission [10.11-10.13], and $g \gg \kappa, \gamma/2$, which is where "vacuum" Rabi splitting is observed [10.14-10.16]. The parameters in Figs. 10.1 and 10.2 invert the conditions for cavity-enhanced and -inhibited spontaneous emission, with g larger then κ and smaller than $\gamma/2$, rather than the reverse. Under these conditions the cavity linewidth is altered by a perturbative coupling to the atom instead of the atomic linewidth being altered by coupling the atom to a cavity mode. We are now going to study the source master equation (10.7) under genuine strong coupling conditions; we will see what happens to optical bistability when g is larger than both κ and $\gamma/2$ ($n_{sat} \ll 1$) – the nonperturbative regime of cavity Q.E.D. We will use the quantum trajectory approach to visualize the quantum fluctuations under these conditions.

10.3 Spontaneous dressed-state polarization

Before we illustrate the fluctuations with quantum trajectories we need to understand how the physics is changed in the strong coupling regime, because, in fact, the physics we have learned from the theory of optical bistability is radically altered; moreover, it is altered in a way that we probably would not expect.

From what we know about the semiclassical theory of optical bistability and the general effects of fluctuations, we might expect the bimodal distributions in Fig. 10.1 to simply be reduced to a single "blob." Strong coupling means $n_{sat} \ll 1$, which means the nonlinearity that gives rise to absorptive optical bistability is turned on by a fraction of a photon. Of course the fraction of a photon is only meaningful as an average quantity, and the fluctuations about this average must be very important. A fluctuation on the scale of one quantum makes the difference between an unsaturated atom (lower branch) and a saturated atom (upper branch). Since quantum mechanics tells us that fluctuations are going to occur on this scale, it is hard to believe that any evidence of the two distinct semiclassical states will remain.

There is nothing wrong with this argument. Certainly the quantum fluctuations are going to be very large. But we need to be suspicious of our prediction of what the large fluctuations will do. The prediction that the bimodality will be washed out is based on the picture of a continuous, diffusive wandering of the system from one region of phase space to another (the picture drawn from the standard Fokker-Planck approach). When single quanta are so important we cannot expect a theory based on a diffusive flow to work very well – we need to incorporate quantum mechanical "jumpiness" in some way.

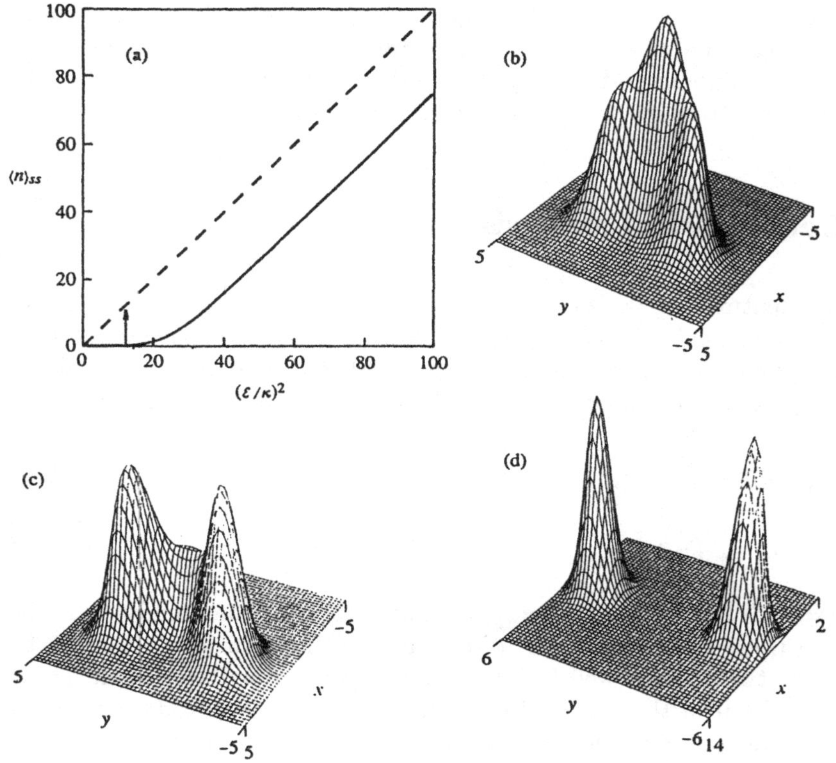

Fig. 10.4. Steady state solution to (10.7) as a function of driving filed intensity for $g/\kappa = 10$ and $\gamma/2\kappa = 0$: (a) mean photon number versus driving field intensity; (b) $Q(x + iy)$ for $\mathcal{E}/\kappa = 4.8$; (c) $Q(x + iy)$ for $\mathcal{E}/\kappa = 5.2$; (d) $Q(x + iy)$ for $\mathcal{E}/\kappa = 10.0$.

What does actually take place is illustrated in Figs. 10.4 and 10.5. These results were obtained by Alsing and Carmichael by numerically solving the master equation (10.7) [10.17]. The figures show the mean photon number as a function of driving field intensity and the Q function for three selected values of intensity. The hysteresis cycle predicted by the semiclassical equation (10.5) is indicated by the vertical arrows in Figs. 10.4(a) and 10.5(a); it consists of the horizontal axis, from the origin out to the vertical arrow, an upwards transition at the arrow, and the return path to the origin along the dashed line (the downwards transition is too small to be resolved). The solid line shows the actual value of the mean photon number which seems to have very little to do with the semiclassical path. The Q functions show just how much the behavior differs from the "washed out bistability" prediction. The bimodality has not just been washed out; it has been replaced by a new bimodality formed from two states separated in the phase direction instead of the amplitude direction. The difference is very clear in Fig. 10.5(b) where the phase and amplitude bimodalities coexist. Note that the phase bimodality persists for arbitrarily large driving field intensities. Alsing and Carmichael call the new bimodality ***spontaneous dressed-state polarization.*** Once we have understood exactly what this is we will be in a position to analyze the fluctuations using quantum trajectories.

10.4 Semiclassical analysis

The main features of the behavior shown in Fig. 10.4 can be understood from a semiclassical calculation, but a different calculation to the one that gave the optical bistability state equation (10.5). The difference comes about by starting from the semiclassical Maxwell-Bloch equations with γ set to zero:

$$\dot{z} = (g/2)v + \mathcal{E} - \kappa z, \tag{10.17a}$$

$$\dot{v} = 2gmz, \tag{10.17b}$$

$$\dot{m} = -g(z^* + v^* z). \tag{10.17c}$$

Here $z = e^{i\omega_C t}\langle a\rangle$, $v = e^{i\omega_C t}\langle\sigma_-\rangle$, and $m = 2\langle\sigma_z\rangle$. The steady-state solutions to these equations are not the same as the solutions obtained by first solving the full Maxwell-Bloch equations (with $\gamma \neq 0$) and then taking the limit $\gamma \to 0$ in the result. Taking the $\gamma \to 0$ limit in different orders gives different answers because a nonzero γ breaks the conservation law $|v|^2 + m^2 = 1$ satisfied by (10.17a)–(10.17c). We do not have time for too many details here. They can be found in [10.17]. The important point is that the steady-state solutions to (10.17a)–(10.17c) bifurcate as a function of the driving field strength at $2\mathcal{E}/g = 1$. For $2\mathcal{E}/g < 1$ there is one stable solution, with

$$z_{ss} = 0, \tag{10.18a}$$

$$v_{ss} = -2\mathcal{E}/g, \tag{10.18b}$$

$$m_{ss} = -\sqrt{1-(2\mathcal{E}/g)^2}. \tag{10.18c}$$

For $2\mathcal{E}/g > 1$ there are two solutions (we will discuss their stability shortly) with

$$z_{ss} = (\mathcal{E}/\kappa)[1-(g/2\mathcal{E})^2] \pm i(g/2\kappa)\sqrt{1-(g/2\mathcal{E})^2}, \tag{10.19a}$$

$$v_{ss} = -(g/2\mathcal{E}) \pm i\sqrt{1-(g/2\mathcal{E})^2}, \tag{10.19b}$$

$$m_{ss} = 0. \tag{10.19c}$$

A plot of $|z_{ss}|^2$ versus $(\mathcal{E}/\kappa)^2$ closely matches the solid curve in Fig. 10.4(a). Also, the locations of the peaks in Fig. 10.4(b) are given by (10.19a).

This bifurcation is completely different from the familiar bifurcation that produces optical bistability. Note, however, that it is not structurally stable, in the sense that for any $\gamma \neq 0$, no matter how small, the solutions (10.18)

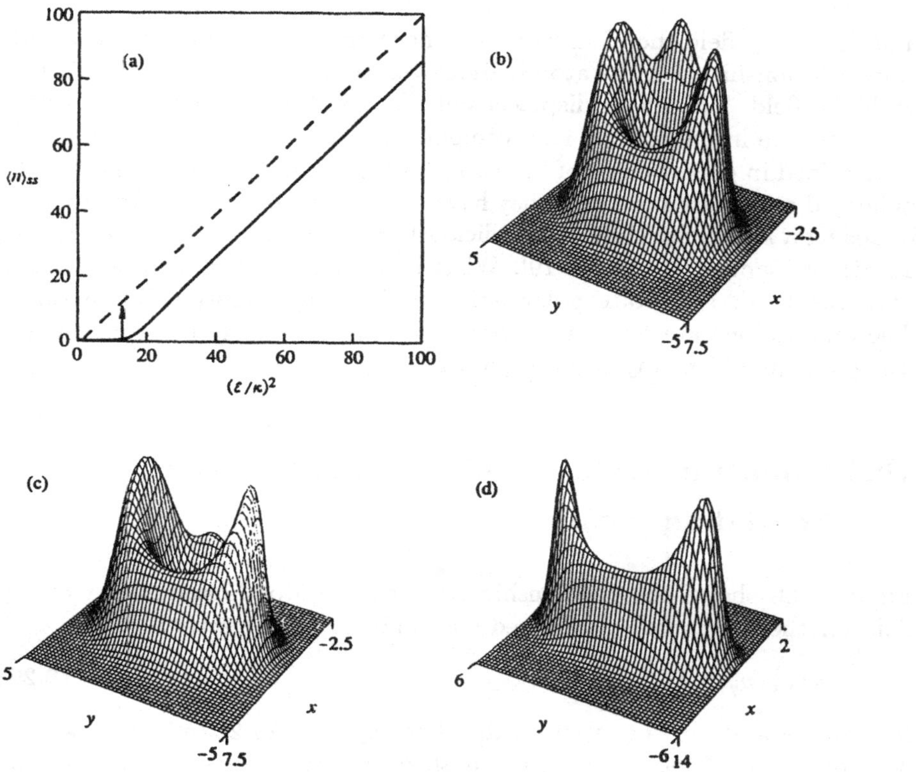

Fig. 10.5. Steady state solution to (10.7) as a function of driving filed intensity for $g/\kappa = 10$ and $\gamma/2\kappa = 1$: (a) mean photon number versus driving field intensity; (b) $Q(x + iy)$ for $\mathcal{E}/\kappa = 4.8$; (c) $Q(x + iy)$ for $\mathcal{E}/\kappa = 5.0$; (d) $Q(x + iy)$ for $\mathcal{E}/\kappa = 10.0$.

and (10.19) are no longer steady-state solutions to the Maxwell-Bloch equations. But when γ is small they are long-lived states, and in the presence of large fluctuations such states will be visited regularly, for relatively long periods of time. Thus, this semiclassical picture makes Fig. 10.4 believable; although, as we will see shortly, it cannot really explain everything when we think a little harder about the fluctuations.

What we get from the semiclassical analysis are clues about the basic physics involved. The most important clue is contained in the results (10.19b) and (10.19c) for the state of the atom. These are the Bloch components for dressed atomic states – states that are stationary in the presence of a resonant classical driving field with complex amplitude (10.19a). Note that the field amplitude (10.19a) and the polarization amplitude (10.19b) both have a component in quadrature to the driving field $\mathcal{E}/\kappa$. Thus, the bifurcation is a symmetry breaking transition: the atom aligns its polarization in one of the dressed states; in so doing it must rotate its phase, and it then radiates an in-quadrature component into the cavity field; the atom and the cavity field therefore work together to find a self-consistent dressed-state relationship with the atomic Bloch vector either aligned or antialigned with the field. The phase displacement seen in Figs. 10.4 and 10.5 is produced by the in-quadrature field components radiated by the atom when it is polarized in one or other of the two possible dressed states. Dressed-state polarized atoms are not new. They have been produced in the laboratory by imposing a $\pi/2$ phase shift, at a judiciously chosen time, on the field driving an atomic sample [10.18, 10.19]. What is different here is that we have a *spontaneous* dressed-state polarization initiated by quantum fluctuations. The fluctuations are our main interest. We are now ready to explain them using the quantum trajectory approach.

10.5 Quantum stability, phase switching, and Schrödinger cats

Figure 10.6 shows the relationship between the atomic states and cavity field for the self-consistent dressed states (10.19). The vector

$$\vec{\sigma} = (v_x, v_y, m) \tag{10.20}$$

locates the state of the atom on the Bloch sphere. As $2\mathcal{E}/g$ increases from zero to unity, $\vec{\sigma}$ moves along the dashed line from the south pole to the equator. For $2\mathcal{E}/g > 1$ there are two possible self-consistent dressed states denoted $\vec{\sigma}_u$ and $\vec{\sigma}_l$. With increasing strength of the driving field these states rotate in opposite directions around the equator so that in the strong driving-field-limit they point in the $+v_y$ and $-v_y$ directions; in this limit $\vec{\sigma}_u$ and $\vec{\sigma}_l$ correspond to the orthogonal dressed states

$$|u\rangle = (1/\sqrt{2})(|+\rangle + i|-\rangle), \tag{10.21a}$$

$$|l\rangle = (1/\sqrt{2})(|+\rangle - i|-\rangle). \tag{10.21b}$$

The vectors $\boldsymbol{B}_u$ and $-\boldsymbol{B}_l$ in Fig. 10.6 are determined by the solutions (10.19a) for the cavity field using the usual magnetic analogy:

$$\boldsymbol{B} = (-2gz_y, 2gz_x, 0). \tag{10.22}$$

The limitations of the semiclassical analysis becomes apparent when we investigate the stability of the solutions (10.19). These solutions are not, in fact, stable, even for $\gamma = 0$. If we consider the dynamics on the Bloch sphere (there is an accompanying motion for z), the two steady states are non-stable fixed points each surrounded by a family of non-stable periodic orbits. A perturbation from one of the steady states just moves the atomic state onto one of the orbits; a further perturbation just moves the state from one orbit to another. In the strong-driving-field limit the periodic orbits are easy to construct and are just circles around the Bloch sphere lying

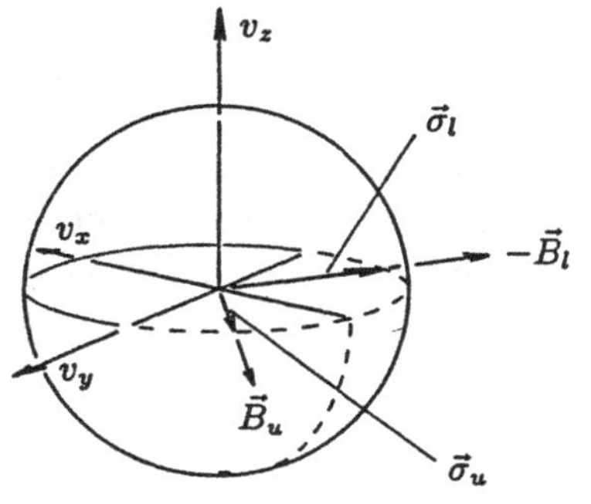

Fig. 10.6. Bloch sphere representation of the self-consistentdressed states (10.19).

in planes perpendicular to the v_y axis (normal undamped Rabi oscillations). The periodic orbits can also be constructed in the bad cavity limit $g/\kappa \ll 1$; here, after adiabatically eliminating the field variable z, the Maxwell-Bloch equations (10.17) are equivalent to the Bloch equations for cooperative resonance fluorescence for which the periodic orbits are known [10.20, 10.21].

The lack of semiclassical stability is important when we consider fluctuations. It means that the standard diffusive picture for the fluctuations leads us to expect that $\vec{\sigma}$ will wander over the entire Bloch sphere. Indeed this is exactly what happens in cooperative resonance fluorescence [10.21]. Now the in-quadrature component of the field is proportional to v_y, and in Fig. 10.4(d) the distribution of this field component is well localized at the two values determined by the $\vec{\sigma}_u$ and $\vec{\sigma}_l$ directions on the Bloch sphere. Random wandering over the Bloch sphere would produce a field distribution stretching continuously between the two peaks of Fig. 10.4(d). Why does this not happen? Where does the stability come from?

We should first consider why the diffusive picture for the fluctuations is inappropriate. The Bloch sphere in Fig. 10.6 has the dimensions of one quantum; the absorption or emission of one photon causes a jump across its diameter. Thus, diffusion across the sphere is just not the right picture if single quantum jumps like this are going to occur. Contrast this situation with the problem of cooperative resonance fluorescence where a diffusive model for the fluctuations does work [10.21]. In that case the Bloch sphere represents the collective pseudo-spin of $N \gg 1$ atoms. It then takes N quantum jumps to cross the Bloch sphere's diameter. On such a sphere, motion generated by many single jumps is accurately represented by diffusion.

We can now see where the stability of our solutions comes from. It is tied to the need for an evolution by quantum jumps; it is the same *quantum stability* that stops the electron spiraling in towards the nucleus in a hydrogen atom. A quantum system can only occupy certain quantized stationary states. In our example the atom has two such states; in the strong-driving-field limit these are the dressed states (10.21). The continuum of intermediate states presumed by a diffusive evolution simply does not exist. Of course, there can be a continuous evolution between stationary states in the sense allowed by superpositions. But dissipative evolution is not of this type. Quantum-mechanical dissipation "jumps." Quantum trajectories provide a way for us to follow the jumps and the coherent evolution between the jumps. [We should really qualify all of this. The jumpy evolution envisages an unravelling of the quantum dynamics that can follow the jumps. The unravelling defined by (10.10)–(10.12) does this. If, however, we used an unravelling based on homodyne detection, like the one in Lecture 9, we would recover a diffusive evolution; albeit a diffusing wavefunction rather than a diffusing phase-space trajectory.]

There is a great deal that could be said about the quantum trajectory treatment of fluctuations for our system. We only have time for a brief overview. To get us started Fig. 10.7 shows three sample trajectories generated by the unravelling (10.10)–(10.12) for the parameters of Fig. 10.5. The figures plot the evolution of the conditioned mean photon number; when time averaged they reproduce the photon number averages read from Fig. 10.5(a). Notice the qualitative change in the character of the fluctuations moving from Fig. 10.7(a) to Fig. 10.7(c). In Fig. 10.7(a) individual quantum transitions associated with the emission of one photon are resolved. This is the regime in which photon antibunching and related nonclassical effects are observed in the field radiated by the cavity [10.4, 10.22-10.24]. Figure 10.7(b) shows a sample quantum trajectory in the threshold region, where the spontaneous dressed-state polarization is trying to get established. The fluctuations are now large and more classical like; although, there is still an occasional return to a state near the vacuum where individual emissions are discernible. For the driving field strength of Fig. 10.7(c) the dressed-state polarization is well established and the conditioned mean photon number shows something like the photon number fluctuations expected for a coherent state.

In fact, the field state in the strong-driving-field limit is not a coherent state. This can be seen in Fig. 10.5(d), which at best represents an ensemble of coherent states with large phase fluctuations. The phase fluctuations seem to span the space separating the coherent states in Fig. 10.4(d). The quantum trajectory approach provides a simple explanation for these phase fluctuations. Imagine that the atom is polarized in the state $|u\rangle$ [Eq. (10.21a)]. The corresponding steady-state field is a coherent state with complex amplitude (10.19a), taken with the positive sign. This is the left-hand peak in Fig. 10.4(d). Now the atom spontaneously emits a photon out the sides of the cavity. The photon frequency will fall within the central peak or the upper sideband of the Mollow spectrum. If it falls within the central peak the atom remains in the state $|u\rangle$. If it falls within the upper sideband the atom makes a transition from the dressed state $|u\rangle$ to the dressed state $|l\rangle$. In this case the steady-state field corresponding to the new atomic state is the coherent state represented by the right-hand peak in Fig. 10.4(d). But the cavity field is not in this state. To get there it must change its phase. The phase of the cavity field therefore begins to switch, driven by

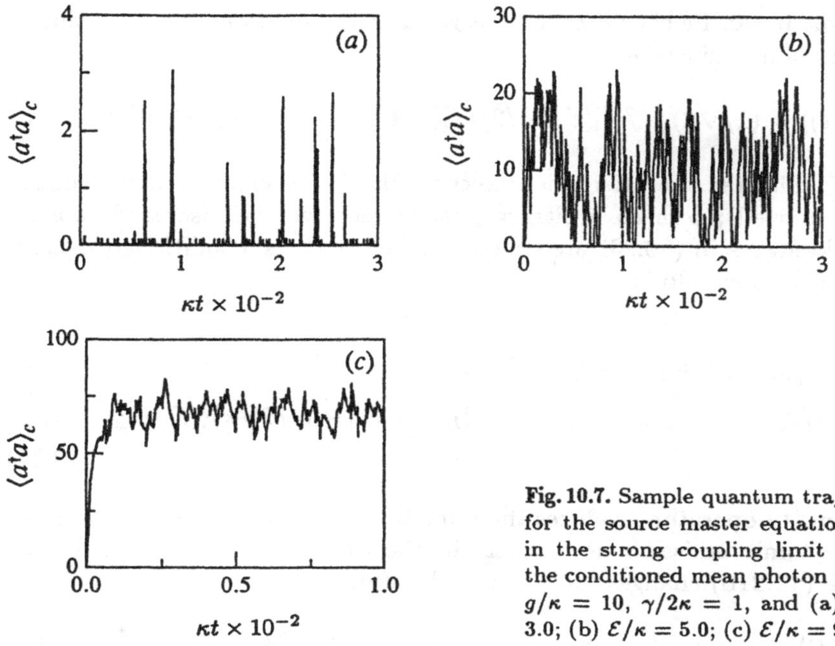

Fig. 10.7. Sample quantum trajectories for the source master equation (10.7) in the strong coupling limit showing the conditioned mean photon number: $g/\kappa = 10$, $\gamma/2\kappa = 1$, and (a) $\mathcal{E}/\kappa = 3.0$; (b) $\mathcal{E}/\kappa = 5.0$; (c) $\mathcal{E}/\kappa = 9.0$.

the changed in-quadrature component of the atomic polarization. Thus, the basic dynamic of the quantum fluctuations in the strong-driving-field-limit is a *phase switching* initiated by individual spontaneous emissions from the atom. Along a quantum trajectory the conditioned Q function will sweep back and forth between the two extremes shown by Fig. 10.4(d) under the direction of the atomic emissions. When the atomic emissions are rare on the time-scale needed for the field to switch its phase ($\sim \kappa^{-1}$) the time averaged Q function shows two distinct peaks [Fig. 10.4(d)]. When the atomic emissions are more frequent, they often catch the field state in midflight, while it is still switching its phase; then the time averaged Q function begins to fill in along the path connecting the peaks [Fig. 10.5(d)]. This picture is given rigorous expression using a unravelling of the master equation designed to visualize the spontaneous transitions between dressed states. The details are worked out in Sect. 5 of [10.17].

The unravelling (10.10)–(10.12) is not quite the same as the one just described because the atomic state collapse (10.12b) does not distinguish between photon emissions into the different peaks of the Mollow spectrum. To conclude this lecture we look briefly at the phase switching generated by (10.10)–(10.12). We will assume that $\gamma/2\kappa \ll 1$, so that the probability of a phase switch being interrupted by an atomic emission before it is complete is small. Consider now some time t during a quantum trajectory, and assume that the last atomic emission occurred many cavity lifetimes ago; the phase switch initiated by the last emission is therefore over and the conditioned wavefuntion has evolved to a temporary steady state. In the strong-driving-field limit it can be shown that, to a good approximation, the conditioned wavefunction is given by

$$|\tilde{\psi}_c(t)\rangle = (1/\sqrt{2})\Big[e^{i\phi/2}|(\mathcal{E}+ig/2)/\kappa\rangle|u\rangle + e^{-i\phi/2}|(\mathcal{E}-ig/2)/\kappa\rangle|l\rangle\Big], \tag{10.23}$$

where $\tilde{}$ indicates the interaction picture (the free oscillation at frequency ω_C is removed); ϕ is an arbitrary phase which will be discussed shortly. The field states in (10.23) are coherent states and the conditioned density operator for the field is

$$\begin{aligned}
&\tilde{\rho}_c^f(t)\\
&= |\langle u|\psi_c(t)\rangle|^2 + |\langle l|\tilde{\psi}_c(t)\rangle|^2\\
&= (1/2)\Big[|(\mathcal{E}+ig/2)/\kappa\rangle\langle(\mathcal{E}+ig/2)/\kappa| + |(\mathcal{E}-ig/2)/\kappa\rangle\langle(\mathcal{E}-ig/2)/\kappa|\Big].
\end{aligned} \tag{10.24}$$

This density operator produces the bimodal Q function in Fig. 10.4(d).

An atomic emission now occurs in the time interval $(t, t+\Delta t]$. The collapse (10.21b) changes (10.23) into the state

$$|\tilde{\psi}_c(t)\rangle = |-\rangle(1/\sqrt{2})\Big[e^{i\phi/2}|(\mathcal{E}+ig/2)/\kappa\rangle + e^{-i\phi/2}|(\mathcal{E}-ig/2)/\kappa\rangle\Big], \tag{10.25a}$$

or, alternatively,

$$\begin{aligned}
|\tilde{\psi}_c(t)\rangle = &-i\frac{1}{2}|u\rangle\Big[e^{i\phi/2}|(\mathcal{E}+ig/2)/\kappa\rangle + e^{-i\phi/2}|(\mathcal{E}-ig/2)/\kappa\rangle\Big]\\
&+ i\frac{1}{2}|l\rangle\Big[e^{i\phi/2}|(\mathcal{E}+ig/2)/\kappa\rangle + e^{-i\phi/2}|(\mathcal{E}-ig/2)/\kappa\rangle\Big].
\end{aligned} \tag{10.25b}$$

The collapsed state (10.25b) contains four terms. Two of the terms involve the product of an atomic dressed state, $|u\rangle$ or $|l\rangle$, and the field state that is stationary for that dressed state – $|\mathcal{E}+ig/2)/\kappa\rangle$ and $|\mathcal{E}-ig/2)/\kappa\rangle$, respectively. These terms are produced by emissions into the central peak of the Mollow spectrum. The other two terms involve products of dressed states with field states that have their phases reversed. These are produced by emissions into the sidebands of the Mollow spectrum. In the subsequent coherent evolution the first two components of the collapsed state will not evolve (except for normalization effects) while the other two components undergo a phase switch. Thus, the Q function splits into four peaks. Two of the peaks sweep through each other as they undergo a phase switch, and reassemble with the other two peaks at the end of the phase switch. This evolution is shown in Fig. 10.8.

It is apparent from this example that the quantum trajectory approach uncovers a lot of detailed dynamics that remain hidden when we simply calculate the steady-state solution to an operator master equation. These dynamics describe the ergodic fluctuations of an individual quantum source. One particularly interesting result revealed for the source we have considered

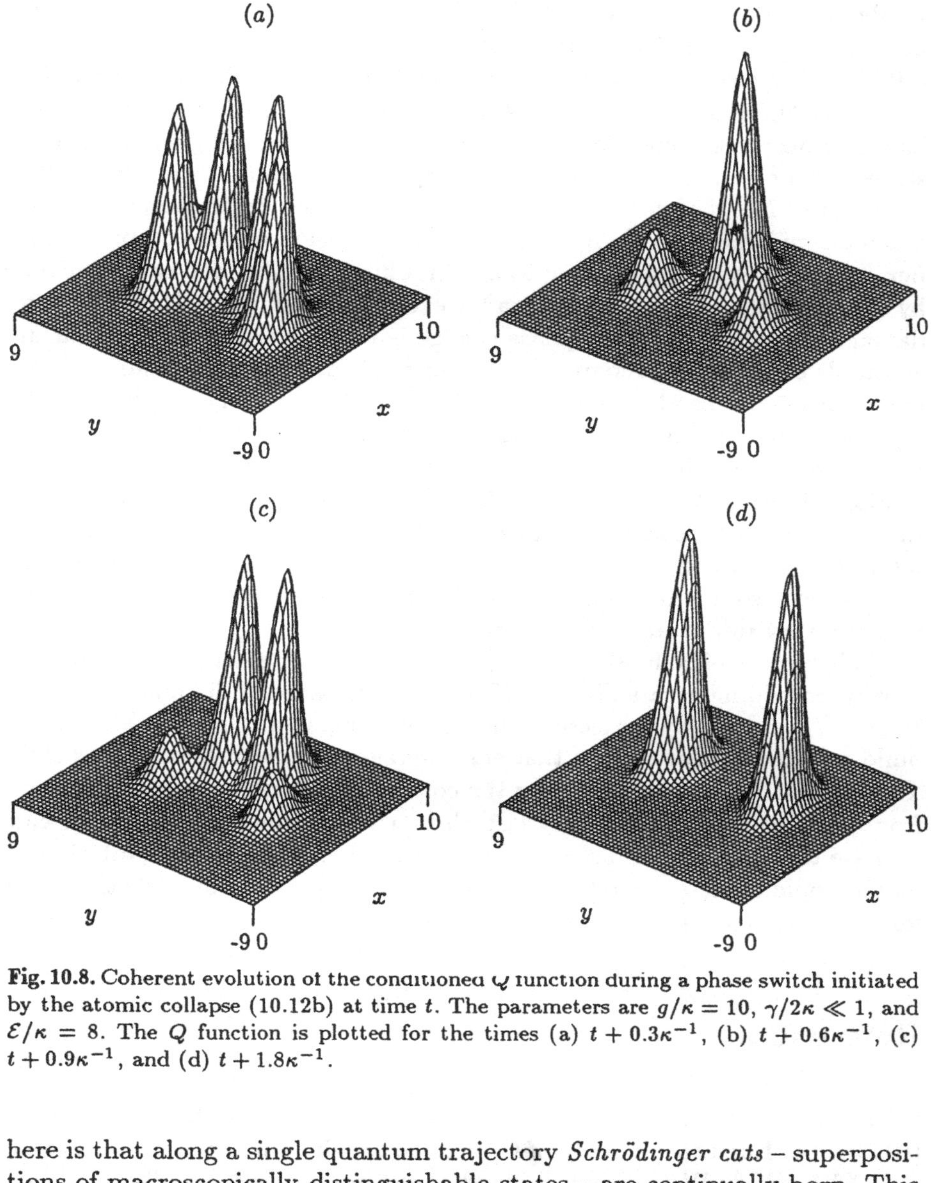

Fig. 10.8. Coherent evolution of the conditioned Q function during a phase switch initiated by the atomic collapse (10.12b) at time t. The parameters are $g/\kappa = 10$, $\gamma/2\kappa \ll 1$, and $\mathcal{E}/\kappa = 8$. The Q function is plotted for the times (a) $t + 0.3\kappa^{-1}$, (b) $t + 0.6\kappa^{-1}$, (c) $t + 0.9\kappa^{-1}$, and (d) $t + 1.8\kappa^{-1}$.

here is that along a single quantum trajectory *Schrödinger cats* – superpositions of macroscopically distinguishable states – are continually born. This follows from (10.25a), which gives a field state that is a superposition of the two nonoverlapping (in phase space) coherent states. It appears that if we do not distinguish between emissions into the different peaks of the Mollow spectrum, each atomic emission gives birth to a Schrödinger cat; the cat turns back into the mixture (10.24) at the end of the phase switch. Is it possible to perform an experiment to catch the cats while they are alive?

Perhaps it is. The main obstacle to be overcome is the phase ϕ that appears in (10.23) and (10.25). This phase depends in a sensitive way on the whole history of coherent evolution and collapse leading up to the time t. To illustrate this we might consider the effect of cavity emissions on the state (10.25a). We have said nothing about cavity emissions, and they certainly take place throughout the evolution illustrated in Fig. 10.8. The reason they have not been mentioned is that we have always been dealing with coherent states. The collapse (10.12a) does nothing to a coherent state. But is that really true? No, it is not quite true. The collapse that accompanies a cavity emission multiplies a coherent state $|\alpha\rangle$ by the phase of the complex number α. This can be devastating to a Schrödinger cat. For the cat described by (10.25a), the phase ϕ changes after each cavity emission. Nothing more damaging happens; but the phase change is bad enough. It means that an ensemble of similarly prepared Schrödinger cats are not really equivalent to one another unless they have suffered an identical history of phase shifting collapses. This is easy to ensure when there are no collapses at all – when there is no dissipation. But, when collapses do occur they occur at random times and in random numbers. Unless countermeasures are taken the ensemble will be randomly phased. The ensemble average then kills the interference terms that tell us the cat is, in fact, a cat. Cavity emissions are not the only source of changes in ϕ. But they illustrate the point that this phase is very important, and difficult to control.

This picture of collapse induced phase-shifts provides a novel explanation of why Schrödinger cats die so swiftly in the presence of dissipation [10.25, 10.26]. Trajectory by trajectory they do not die at all. The problem is to build a measurement scheme that only averages an ensemble of *phased* cats. To do this, ideally we must know the complete history of photon emissions from the cat, down to the very last photon. With this information we can rephase the ensemble and see the cat. But we do not follow the evolution of macroscopic objects down to the level of every quantum jump. We therefore miss the cats that stalk the world of individual quantum trajectories.

References

[10.1] G. P. Agrawal and H. J. Carmichael, *Phys. Rev.* A **19**, 2074 (1979).

[10.2] L. A. Lugiato, "Theory of Optical Bistability," in *Progress in Optics*, Vol. XXI, ed. E. Wolf, North Holland: Amsterdam, 1984, pp. 69ff.

[10.3] H. J. Carmichael, "Quantum Fluctuations in Optical Bistability," in *Frontiers in Quantum Optics*, eds. E. R. Pike and S. Sarkar, Adam Hilger: Bristol, 1986, pp. 120ff.

[10.4] F. Casagrande and L. A. Lugiato, *Nuovo Cim.* B **55**, 173 (1980).

[10.5] L. A. Lugiato and G. Strini, *Optics Commun.* **41**, 67 (1982).

[10.6] P. D. Drummond and C. W. Gardiner, *J. Phys.* A **13**, 2353 (1980).

[10.7] H. J. Carmichael, J. S. Satchell, and S. Sarkar, *Phys. Rev.* A **34**, 3166 (1986).

[10.8] M. Dörfle and A. Schenzle, *Z. Phys.* B **65**, 113 (1986).
[10.9] A. M. Smith and C. W. Gardiner, *Phys. Rev.* A **39**, 3511 (1989).
[10.10] C. M. Savage and H. J. Carmichael, *IEEE J. Quantum Electron.* **24**, 1495 (1988).
[10.11] E. M. Purcell, *Phys. Rev.* **69**, 681 (1946).
[10.12] D. Kleppner, *Phys. Rev. Lett.* **47**, 233 (1981).
[10.13] S. Haroche and J. M. Raimond, in *Advances in Atomic and Molecular Physics*, eds. D. Bates and B. Bederson, Academic Press: New York, 1985, pp. 347ff.
[10.14] J. J. Sanchez-Mondragon, N. B. Narozhny, and J. H. Eberly, *Phys. Rev. Lett.* **51**, 550 (1983).
[10.15] M. G. Raizen, R. J. Thompson, R. J. Brecha, H. J. Kimble, and H. J. Carmichael, *Phys. Rev. Lett.* **63**, 240 (1989).
[10.16] Y. Zhu, D. J. Gauthier, S. E. Morin, Q. Wu, H. J. Carmichael, and T. W. Mossberg, *Phys. Rev. Lett.* **64**, 2499 (1990).
[10.17] P. Alsing and H. J. Carmichael, *Quantum Optics* **3**, 13 (1991).
[10.18] Y. S. Bai, A. G. Yodh, and T. W. Mossberg, *Phys. Rev. Lett.* **55**, 1277 (1985).
[10.19] J. E. Golub, Y. S. Bai, and T. W. Mossberg, *Phys. Rev.* A **37**, 119 (1988).
[10.20] P. D. Drummond and H. J. Carmichael, *Optics Commun.* **27**, 160 (1978).
[10.21] H. J. Carmichael, *J. Phys.* B **13**, 3551 (1980).
[10.22] P. R. Rice and H. J. Carmichael, *IEEE J. Quantum Electron.* **24**, 1351 (1988).
[10.23] H. J. Carmichael, R. J. Brecha, and P. R. Rice, *Optics Commun.* **82**, 73 (1991).
[10.24] G. Rempe, R. J. Thompson, R. J. Brecha, W. D. Lee, and H. J. Kimble, *Phys. Rev. Lett.* **67**, 1727 (1991).
[10.25] A. O. Caldeira and A. J. Leggett, *Phys. Rev.* A **31**, 1059 (1985).
[10.26] D. F. Walls and M. G. Milburn, *Phys. Rev.* A **31**, 2103 (1985).

Postscript

Lectures 7–10 describe the quantum trajectory idea from the perspective of my own work. They are based upon the understanding I had of the subject and the related literature at the end of 1991. This postscript is an attempt to set the lectures in a broader context, to provide references to related work that has appeared during the last year and to work I was unaware of in 1991.

Since the referencing in the lectures is a little sparse, let me first say something about the connections between my work and earlier work in quantum optics. The development of quantum trajectory theory presented in the lectures, starting from the photoelectron counting formula 7.1, proceeding to the expression (7.19) for exclusive probability densities, and from there making a connection with a source master equation, follows the evolution of my own thinking on this subject. The basic ideas appear in a paper written with Surendra Singh, Reeta Vyas, and Perry Rice on waiting times and state reduction in resonance fluorescence [1]; although, important developments beyond what is contained in that paper were made to arrive at the general theory outlined in the lectures. As stated in [1], my attention was first turned in the direction leading to quantum trajectories by the quantum jump work of Cohen-Tannoudji and Dalibard [2], and Zoller, Marte, and Walls [3]. This work caused me to look in some detail at the relationship between exclusive and nonexclusive photoelectron counting probabilities – principally because it posed, for me, a puzzle: The message of these authors was essentially that quantum jumps are more easily understood in terms of the waiting-time distribution $w(\tau)$ than the second-order correlation function $g^{(2)}(\tau)$ (Sect. 7.2). For me (due to ignorance) the contrast between the two quantities was a puzzle because I knew that experiments on photon antibunching in resonance fluorescence actually measured waiting-time distributions, and yet the measurements were reported as results for second-order correlation functions [4]. How, then, could the difference between the two be so important? What, in fact, was the difference, and when could it be overlooked? Answering these questions lead me to the rewriting of the standard photoelectron counting formula described in Lecture 7, and to the connection between the counting formula and the source master equation that forms the basis of quantum trajectory theory.

I recognized at the time that the mathematical language of the rewritten photoelectron counting formula was that of Srinivas and Davies [5]; indeed,

Zoller, Marte, and Walls [3] had noted that their equations, based on a theory of resonance fluorescence by Mollow [6], had the mathematical form of the photoelectron counting theory of Srinivas and Davies. It was also clear to me (see [1]) that the use of exclusive probability densities was implicit in Mollow's derivation of the photon counting distribution for resonance fluorescence [6], and in a similar derivation by Cook [7]. What appeared to be missing in all this earlier work, however, was a clear and general statement of the connection between the mathematics of Srinivas and Davies, the standard theory of photoelectric detection (the counting formula of Mandel, Glauber, and Kelly and Kleiner), and the theory of photo-emissive sources (operator master equations). In fact, the work of Srinivas and Davies obscured the connection by suggesting that the standard theory of photoelectric detection is inadequate [4]. Mandel had answered their criticism with a physical explanation of why the standard theory is valid (assuming it is not grossly misapplied) [8,9]. Nevertheless, the Srinivas and Davies theory continued to be quoted in quantum optics circles, independently of standard photoelectron counting theory, for nearly ten years, without the explicit connection between their mathematics and Mandel's physics being made. To my knowledge, the connection was made for the first time in [1] (at the end of section V).

It is this connection, played out in the relationship between source dynamics and photoelectron counting sequences that, as described in the lectures: (1) suggests the formulation of a general theory that goes beyond the special case of direct (gedanken) detection of the radiation from a two- or few-state atom; (2) allows for a systematic interpretation of different quantum trajectories (unravellings) based on different arrangements of measuring apparatus [provides a concrete, in-the-laboratory (not just in-the-imagination) connection to quantum measurement questions]. During the last year I have become aware of a large amount of work that is more or less closely related to quantum trajectory theory [10-49]. In the interest of not delaying this volume further I will not attempt to delineate all the similarities and differences between the ideas found in this literature and my work; nor will I attempt any detailed comparisons amongst the papers in the literature. I do emphasize what I have just said: The principal thing characterizing quantum trajectory theory is the explicit connection it builds between the stochastic wavefunction trajectory and the classical stochastic outputs of detectors that monitor the system the wavefunction describes. In addition, it is an essential feature that the connection is not dogmatic, but has a flexible form that depends on the arrangement of the detection scheme (direct versus homodyne detection for example). There is certainly some overlap with these ideas in some of the papers listed below [10-49]. Nevertheless, to my mind, none of them works out the physical basis of the source-wavefunction-detected-signal connection in such a complete and systematic way.

My reference list is definitely incomplete. I only have to backtrack a short way through the literature referenced in a few of the quoted papers to double or triple the length of the list. The order of the references is primarily determined by the order in which preprints and reprints have piled up on my desk.

References [10-19] are very closely connected to quantum trajectory theory. More specifically, they concern the direct detection unravelling of a source master equation (Lectures 7 and 8). They are the result of independent constructions of a stochastic wavefunction evolution equivalent to the master equation for a radiating atom by Dalibard et al. [10] and Hegerfeldt and Wilser [16]. The work of Zoller et al. [12-15], while it received some stimulus [12] from discussions with Dalibard, is developed in the language of Mollow [6] and Srinivas and Davies [5], and in this sense is quite independent of [10]. The Srinivas and Davies language is also applied extensively in work by Ueda et al. [20-27]. The more formal parts of the work of Zoller et al. [14] draw on the methods of quantum stochastic calculus used by some of the other authors, especially Barchielli [28-30] and Belavkin [31–33]. Setting aside the formality and different starting point of Barchielli's work, there is, in one sense, more overlap with quantum trajectories here than elsewhere. Barchielli considers homodyne (and heterodyne) detection in addition to direct detection, which is not done by the other quantum optics authors. The homodyne detection unravelling (Lecture 9) is also connected with work by Gisin [34-38]. Gisin starts from a quite different position, constructing a stochastic wavefunction evolution on the basis of formal measurement theory arguments. Nevertheless, it is clear that the continuous, nonlinear, stochastic equations he considers are of essentially the same mathematical type as the homodyne detection unravelling [Eqs. (9.29)–(9.31)]. Some of the simulations he has performed recently with Percival [36] are very similar to simulations we have obtained from our homodyne detection equations (not in the lectures). During the last year the quantum trajectory idea has been filled out and compared with some of the alternative approaches by Wiseman and Milburn [39-41].

The list of references is already quite diverse and demonstrates a strong convergence of ideas on the use of stochastic wavefunctions in quantum mechanics. The connections, however, are still broader. The themes in the references mentioned so far are principally: (1) radiating (open) systems in quantum optics; (2) quantum measurement – particular of the continuous sort encountered in quantum optics. One other theme that impinges on the quantum trajectory idea must be mentioned. It concerns two related issues: (1) What form should the fundamental dynamical equations of physics take? Are they to be based on a unitary evolution? If so, how do we extract the open system description used in the lectures from the more fundamental unitary description? (2) How are the quantum states of an unstable system (particle) to be defined? These issues involve the long-standing question of irreversibility, and, more specifically, the central role that irreversibility plays when we try to understand quantum mechanics. The lectures do not attempt to reach the philosophical and mathematical consistency on these questions that one would hope to build into a fundamental theory. In fact, the difficulties are glossed over – in two places: first, when the master equation for a photoemissive source is derived (Lectures 1), where the Born-Markoff approximation is invoked without apology; second, when the photoelectric detector is simply presented, ready made, as a device that outputs a classical stochastic counting process (photoelectron sequences) from an input of quantized fields [Eq. (7.1)] – here the break is made on the basis of perturbation theory. The implicit assertion is, of course, that the final quantum trajectory description is physically "correct," and somewhere close to the place that must be reach after the philosophical and mathematical niceties are more convincingly addressed.

There is a vast literature on irreversibility and its connection to the fundamental equations of physics. I will give only a few references that are related closely to quantum trajectories. The question concerning states for unstable systems arises in a prominent way in particle physics. There one deals constantly with objects whose existence, in human terms, is transient in the extreme. Sudarshan has a long-standing interest in the question [42-44]; the operator master equations that describe photoemissive sources will be found in his work on dynamical semigroups [45,46]. The issues of irreversibility and quantum measurement are also currently being addressed in connection with the evolution of the ultimate closed system – the universe as a whole. The work by Gell-Mann and Hartle on this subject is widely known [47-49]. The dynamical evolution reached via "decoherence" in their theory is very similar to the evolution of a quantum trajectory; although the grounding in closed, rather than open system dynamics is a fundamental distinction.

In the year that has passed since I presented the ULB Lectures my students and I have continued to work on quantum trajectories, applying the ideas to the spectroscopy of a cavity Q.E.D. system [50] and to the generation and detection of optical Schrödinger cats [51]. I have extended the ideas in a fundamental way by working out the basic principles of the quantum trajectory theory for cascaded open systems [52]. Crispin Gardiner has also addressed this problem; but without using the language of quantum trajectories [53].

I hope, in the near future, to find time to explore the literature referenced here in more depth. It seems clear that there is common thinking on the subjects of irreversibility and quantum measurement taking place across a broad range of research areas. I look forward to seeing what new understanding will be refined from all this work.

References

[1] H. J. Carmichael, S. Singh, R. Vyas, and P. R. Rice, *Phys. Rev.* A **39**, 1200 (1989).
[2] C. Cohen-Tannoudji and J. Dalibard, *Europhys. Lett.* **1**, 441 (1986).
[3] P. Zoller, M. Marte, and D. F. Walls, *Phys. Rev.* A **35**, 198 (1987).
[4] H. J. Kimble, M. Dagenais, and L. Mandel, *Phys. Rev. Lett.* **39**, 691 (1977).
[5] M. D. Srinivas and E. B. Davies, *Optica Acta*, **28**, 981 (1981).
[6] B. R. Mollow, *Phys. Rev.* A **12**, 1919 (1975).
[7] R. J. Cook, *Phys. Rev.* A **23**, 1243 (1981).
[8] L. Mandel, *Optica Acta* **28**, 1447 (1981).
[9] M. D. Srinivas and E. B. Davies, *Optica Acta* **29**, 235 (1982).
[10] J. Dalibard, Y. Castin, and K. Mølmer, *Phys. Rev. Lett.* **68**, 580 (1992).
[11] K. Mølmer, Y. Castin, and J. Dalibard, *J. Opt. Soc. Am.* B, in press.
[12] R. Dum, P. Zoller, and H. Ritsch, *Phys. Rev.* A **45**, 4879 (1992).
[13] R. Dum, A. S. Parkins, P. Zoller, and C. W. Gardiner, *Phys. Rev.* A **46**, 4382 (1992).
[14] C. W. Gardiner, A. S. Parkins, and P. Zoller, *Phys. Rev.* A **46**, 4363 (1992).
[15] P. Marte, R. Dum, R. Taïeb, and P. Zoller, Comment on "Observation of quantized motion of Rb atoms in an optical field", preprint.
[16] G. C. Hegerfeldt and T. S. Wilser, in *Proceedings of the II International Wigner Symposium*, Goslar, Germany, July 1991, eds. H. D. Doebner, W. Scherer, and F. Schroeck, World Scientific: Singapore, 1992.
[17] G. C. Hegerfeldt and M. B. Plenio, *Phys. Rev.* A **46**, 373 (1992).
[18] G. C. Hegerfeldt and M. B. Plenio, "Coherence with incoherent light: A new type of quantum beat for a single atom," preprint.
[19] G. C. Hegerfeldt, "How to reset an atom after a photon detection: Applications to photon counting processes," preprint.
[20] M. Ueda, *Quantum Opt.* **1**, 131 (1989).
[21] M. Ueda, N. Imoto, and T. Ogawa, *Phys. Rev.* A **41**, 3891 (1990).
[22] N. Imoto, M. Ueda, and T. Ogawa, *Phys. Rev.* A **41**, 4127 (1990).
[23] M. Ueda, *Phys. Rev.* A **41**, 3875 (1990).
[24] M. Ueda, N. Imoto, T. Ogawa, *Phys. Rev.* A **41**, 6331 (1990).
[25] T. Ogawa, M. Ueda, and N. Imoto, *Phys. Rev. Lett.* **66**, 1046 (1991).
[26] T. Ogawa, M. Ueda, and N. Imoto, *Phys. Rev.* A **43**, 6458 (1991).
[27] M. Ueda and M. Kitagawa, *Phys. Rev. Lett.* **68**, 3424 (1992).
[28] A. Barchielli, *Quantum Opt.* **2**, 423 (1990).
[29] A. Barchielli and V. P. Belavkin, *J. Phys.* A **24**, 1495 (1991).
[30] A. Barchielli, "Stochastic differential equations and 'a posteriori' states in quantum mechanics," preprint.
[31] V. P. Belavkin, *J. Phys.* A **22**, L1109 (1989).
[32] V. P. Belavkin and P. Staszewski, *Phys. Rev.* A **45**, 1347 (1992).

[33] P. Staszewski and G. Staszewska, *Open Systems and Information Dynamics* **1**, 103 (1992).
[34] N. Gisin, *Phys. Rev. Lett.* **52**, 1657 (1984).
[35] N. Gisin and M. B. Cibils, *J. Phys.* A **25**, 5165 (1992).
[36] N. Gisin and I. C. Percival, *J. Phys.* A **25**, 5677 (1992).
[37] N. Gisin and I. C. Percival, "Quantum state diffusion, localisation and quantum dispersion entropy," preprint.
[38] N. Gisin and I. C. Percival, "The quantum state diffusion picture of physical processes," preprint.
[39] H. M. Wiseman and G. J. Milburn, *Phys. Rev.* A **47**, 642 (1993).
[40] H. M. Wiseman and G. J. Milburn, "The interpretation of quantum jump and diffusion processes illustrated on the Bloch sphere," preprint.
[41] H. M. Wiseman and G. J. Milburn, "Quantum theory of optical feedback via homodyne detection," preprint.
[42] P. M. Mathews, J. Rau, and E. C. G. Sudarshan, *Phys. Rev.* **121**, 920 (1961).
[43] E. C. G. Sudarshan, C. B. Chiu, and V. Gorini, *Phys. Rev.* D **18**, 2914 (1978).
[44] C. B. Chiu and E. C. G. Sudarshan, *Phys. Rev.* D **42**, 3712 (1990).
[45] E. C. G. Sudarshan, "Quantum dynamics, metastable states and contractive semigroups," University of Texas preprint DOE-ER40200-265 (1991).
[46] E. C. G. Sudarshan, "The structure of quantum dynamical semigroups," University of Texas preprint DOE-ER40200-270 (1991).
[47] M. Gell-Mann and J. B. Hartle, in *Complexity, Entropy, and the Physics of Information, SFI Studies in the Science of Complexity*, Vol. III. ed. W. Zurek, Addison Wesley: Reading, 1990.
[48] M. Gell-Mann and J. B. Hartle, "Alternative decohering histories in quantum mechanics," in *Proceedings of the 25th International Conference on High Energy Physics*, Singapore, August 2-8, 1990, eds. K. K. Phua and Y. Yamaguchi, World Scientific: Singapore, 1990.
[49] M. Gell-Mann and J. B. Hartle, "Classical equations for quantum systems," Caltech preprint CALT-68-1834 (1991).
[50] L. Tian and H. J. Carmichael, *Phys. Rev.* A **46**, R6801 (1992).
[51] H. J. Carmichael, L. Tian, W. Ren, and P. Alsing, "Nonperturbative atom-photon interactions in an optical cavity," in *Cavity Quantum Electrodynamics*, ed. P. R. Berman, Academic Press: Orlando, in press.
[52] H. J. Carmichael, "Quantum trajectory theory for cascaded open systems," preprint.
[53] C. W. Gardiner, "A quantum system driven by the output field from another quantum system," preprint.

Lecture Notes in Physics

For information about Vols. 1–379
please contact your bookseller or Springer-Verlag

Vol. 380: I. Tuominen, D. Moss, G. Rüdiger (Eds.), The Sun and Cool Stars: activity, magnetism, dynamos. Proceedings, 1990. X, 530 pages. 1991.

Vol. 381: J. Casas-Vazquez, D. Jou (Eds.), Rheological Modelling: Thermodynamical and Statistical Approaches. Proceedings, 1990. VII, 378 pages. 1991.

Vol. 382: V.V. Dodonov, V. I. Man'ko (Eds.), Group Theoretical Methods in Physics. Proceedings, 1990. XVII, 601 pages. 1991.

Vol. 384: M. D. Smooke (Ed.), Reduced Kinetic Mechanisms and Asymptotic Approximations for Methane-Air Flames. V, 245 pages. 1991.

Vol. 385: A. Treves, G. C. Perola, L. Stella (Eds.), Iron Line Diagnostics in X-Ray Sources. Proceedings, Como, Italy 1990. IX, 312 pages. 1991.

Vol. 386: G. Pétré, A. Sanfeld (Eds.), Capillarity Today. Proceedings, Belgium 1990. XI, 384 pages. 1991.

Vol. 387: Y. Uchida, R. C. Canfield, T. Watanabe, E. Hiei (Eds.), Flare Physics in Solar Activity Maximum 22. Proceedings, 1990. X, 360 pages. 1991.

Vol. 388: D. Gough, J. Toomre (Eds.), Challenges to Theories of the Structure of Moderate-Mass Stars. Proceedings, 1990. VII, 414 pages. 1991.

Vol. 389: J. C. Miller, R. F. Haglund (Eds.), Laser Ablation-Mechanisms and Applications. Proceedings. IX, 362 pages, 1991.

Vol. 390: J. Heidmann, M. J. Klein (Eds.), Bioastronomy - The Search for Extraterrestrial Life. Proceedings, 1990. XVII, 413 pages. 1991.

Vol. 391: A. Zdziarski, M. Sikora (Eds.), Ralativistic Hadrons in Cosmic Compact Objects. Proceedings, 1990. XII, 182 pages. 1991.

Vol. 392: J.-D. Fournier, P.-L. Sulem (Eds.), Large-Scale Structures in Nonlinear Physics. Proceedings. VIII, 353 pages. 1991.

Vol. 393: M. Remoissenet, M.Peyrard (Eds.), Nonlinear Coherent Structures in Physics and Biology. Proceedings. XII, 398 pages. 1991.

Vol. 394: M. R. J. Hoch, R. H. Lemmer (Eds.), Low Temperature Physics. Proceedings. X,374 pages. 1991.

Vol. 395: H. E. Trease, M. J. Fritts, W. P. Crowley (Eds.), Advances in the Free-Lagrange Method. Proceedings, 1990. XI, 327 pages. 1991.

Vol. 396: H. Mitter, H. Gausterer (Eds.), Recent Aspects of Quantum Fields. Proceedings. XIII, 332 pages. 1991.

Vol. 398: T. M. M. Verheggen (Ed.), Numerical Methods for the Simulation of Multi-Phase and Complex Flow. Proceedings, 1990. VI, 153 pages. 1992.

Vol. 399: Z. Švestka, B. V. Jackson, M. E. Machedo (Eds.), Eruptive Solar Flares. Proceedings, 1991. XIV, 409 pages. 1992.

Vol. 400: M. Dienes, M. Month, S. Turner (Eds.), Frontiers of Particle Beams: Intensity Limitations. Proceedings, 1990. IX, 610 pages. 1992.

Vol. 401: U. Heber, C. S. Jeffery (Eds.), The Atmospheres of Early-Type Stars. Proceedings, 1991. XIX, 450 pages. 1992.

Vol. 402: L. Boi, D. Flament, J.-M. Salanskis (Eds.), 1830-1930: A Century of Geometry. VIII, 304 pages. 1992.

Vol. 403: E. Balslev (Ed.), Schrödinger Operators. Proceedings, 1991. VIII, 264 pages. 1992.

Vol. 404: R. Schmidt, H. O. Lutz, R. Dreizler (Eds.), Nuclear Physics Concepts in the Study of Atomic Cluster Physics. Proceedings, 1991. XVIII, 363 pages. 1992.

Vol. 405: W. Hollik, R. Rückl, J. Wess (Eds.), Phenomenological Aspects of Supersymmetry. VII, 329 pages. 1992.

Vol. 406: R. Kayser, T. Schramm, L. Nieser (Eds.), Gravitational Lenses. Proceedings, 1991. XXII, 399 pages. 1992.

Vol. 407: P. L. Smith, W. L. Wiese (Eds.), Atomic and Molecular Data for Space Astronomy. VII, 158 pages. 1992.

Vol. 408: V. J. Martínez, M. Portilla, D. Sàez (Eds.), New Insights into the Universe. Proceedings, 1991. XI, 298 pages. 1992.

Vol. 409: H. Gausterer, C. B. Lang (Eds.), Computational Methods in Field Theory. Proceedings, 1992. XII, 274 pages. 1992.

Vol. 410: J. Ehlers, G. Schäfer (Eds.), Relativistic Gravity Research. Proceedings, VIII, 409 pages. 1992.

Vol. 411: W. Dieter Heiss (Ed.), Chaos and Quantum Chaos. Proceedings, XIV, 330 pages. 1992.

Vol. 412: A. W. Clegg, G. E. Nedoluha (Eds.), Astrophysical Masers. Proceedings, 1992. XX, 480 pages. 1993.

Vol. 413: Aa. Sandqvist, T. P. Ray (Eds.); Central Activity in Galaxies. From Observational Data to Astrophysical Diagnostics. XIII, 235 pages. 1993.

Vol. 414: M. Napolitano, F. Sabetta (Eds.), Thirteenth International Conference on Numerical Methods in Fluid Dynamics. Proceedings, 1992. XIV, 541 pages. 1993.

Vol. 415: L. Garrido (Ed.), Complex Fluids. Proceedings, 1992. XIII, 413 pages. 1993.

Vol. 416: B. Baschek, G. Klare, J. Lequeux (Eds.), New Aspects of Magellanic Cloud Research. Proceedings, 1992. XIII, 494 pages. 1993.

Vol. 417: K. Goeke P. Kroll, H.-R. Petry (Eds.), Quark Cluster Dynamics. Proceedings, 1992. XI, 297 pages. 1993.

New Series m: Monographs

Vol. m 1: H. Hora, Plasmas at High Temperature and Density. VIII, 442 pages. 1991.

Vol. m 2: P. Busch, P. J. Lahti, P. Mittelstaedt, The Quantum Theory of Measurement. XIII, 165 pages. 1991.

Vol. m 3: A. Heck, J. M. Perdang (Eds.), Applying Fractals in Astronomy. IX, 210 pages. 1991.

Vol. m 4: R. K. Zeytounian, Mécanique des fluides fondamentale. XV, 615 pages, 1991.

Vol. m 5: R. K. Zeytounian, Meteorological Fluid Dynamics. XI, 346 pages. 1991.

Vol. m 6: N. M. J. Woodhouse, Special Relativity. VIII, 86 pages. 1992.

Vol. m 7: G. Morandi, The Role of Topology in Classical and Quantum Physics. XIII, 239 pages. 1992.

Vol. m 8: D. Funaro, Polynomial Approximation of Differential Equations. X, 305 pages. 1992.

Vol. m 9: M. Namiki, Stochastic Quantization. X, 217 pages. 1992.

Vol. m 10: J. Hoppe, Lectures on Integrable Systems. VII, 111 pages. 1992.

Vol. m 11: A. D. Yaghjian, Relativistic Dynamics of a Charged Sphere. XII, 115 pages. 1992.

Vol. m 12: G. Esposito, Quantum Gravity, Quantum Cosmology and Lorentzian Geometries. XVI, 326 pages. 1992.

Vol. m 13: M. Klein, A. Knauf, Classical Planar Scattering by Coulombic Potentials. V, 142 pages. 1992.

Vol. m 14: A. Lerda, Anyons. XI, 138 pages. 1992.

Vol. m 15: N. Peters, B. Rogg (Eds.), Reduced Kinetic Mechanisms for Applications in Combustion Systems. X, 360 pages. 1993.

Vol. m 16: P. Christe, M. Henkel, Introduction to Conformal Invariance and Its Applications to Critical Phenomena. XV, 260 pages. 1993.

Vol. m 17: M. Schoen, Computer Simulation of Condensed Phases in Complex Geometries. X, 136 pages. 1993.

Vol. m 18: H. Carmichael, An Open Systems Approach to Quantum Optics. X, 179 pages, 1993.

Made in the USA
Monee, IL
13 September 2026

62973759R00109